"These respected Arctic scholars provide valuable insights into the complexity of the timely and important subject of Arctic governance, providing thoughtful discussions of the political, economic, legal, and cultural dimensions of governance at local, subnational, national, and regional levels."

– **Elizabeth Riddell-Dixon**, *Professor Emerita, Western University, Canada*

"*Governing Complexity in the Arctic Region* is a must-read for anyone interested in Arctic issues. It gives the reader an excellent introduction to how different actors have shaped the 'new' Arctic agenda."

– **Dr. Svein Vigeland Rottem**, *Senior Research Fellow, Fridtjof Nansen Institute, Norway*

"As a scholar who has written extensively on the process of providing a framework for governance in the northern circumpolar region, I find that their recent undertaking offers a useful consideration of the issues and problems involved. They are knowledgeable of the Arctic region and its present needs. Their work gets at the heart of the challenges and opportunities that confront those who engage in diplomacy and paradiplomacy within the circumpolar arena."

– **Douglas C. Nord**, *Professor, Umeå University, Sweden*

Governing Complexity in the Arctic Region

This book examines emerging forms of governance in the Arctic region, exploring how different types of state and non-state actors promote and support rules and standards.

The authors argue that confining our understandings of Arctic governance to Arctic states and a focus on the Arctic Council as the primary site of circumpolar governance provides an incomplete picture. Instead, they embrace the complexity of governance in the Arctic by systematically analyzing and comparing the position, interventions, and influence of different actor groups seeking to shape Arctic political and economic outcomes in multiple sites of Arctic politics, both formal and informal. This book assesses the potential that sub-national governments, corporations, civil society organizations, Indigenous peoples, and non-Arctic states possess to develop norms and standards to ensure a stable, rule-based Arctic region. It will be of interest to all scholars and students working in the fields of Arctic Sovereignty, Security Studies, Global Governance, and International Political Economy.

Mathieu Landriault is the director of the Observatoire de la Politique et la Sécurité de L'Arctique (OPSA), based in Montreal.

Andrew Chater is an assistant professor (limited term) at Brescia University College.

Elana Wilson Rowe is Research Professor at the Norwegian Institute of International Affairs (NUPI).

P. Whitney Lackenbauer is Canada Research Chair (Tier 1) in the Study of the Canadian North at Trent University.

The International Political Economy of New Regionalisms Series
Series Editor: Timothy M. Shaw

The International Political Economy of New Regionalisms Series presents innovative analyses of a range of novel regional relations and institutions. Going beyond established, formal, interstate economic organizations, this essential series provides informed interdisciplinary and international research and debate about myriad heterogeneous intermediate-level interactions. Reflective of its cosmopolitan and creative orientation, this series is developed by an international editorial team of established and emerging scholars in both the South and North. It reinforces ongoing networks of analysts in both academia and think-tanks as well as international agencies concerned with micro-, meso- and macro-level regionalisms.

Structural Change in Africa
Misperceptions, New Narratives and Development in the 21st Century
Carlos Lopes and George Kararach

Multipolarization, South-South Cooperation, and the Rise of Post-Hegemonic Governance
Efe Can Gürcan

Inclusive Trade in Africa
The African Continental Free Trade Area in Comparative Perspective
Edited by David Luke and Jamie MacLeod

The Political Economy of New Regionalisms in the Pacific Rim
Edited by José Briceño Ruiz and Philippe De Lombaerde

Governing Complexity in the Arctic Region
Mathieu Landriault, Andrew Chater, Elana Wilson Rowe and P. Whitney Lackenbauer

For more information about this series, please visit: www.routledge.com/ASHSER-1146

Governing Complexity in the Arctic Region

Mathieu Landriault, Andrew Chater, Elana Wilson Rowe and P. Whitney Lackenbauer

Routledge
Taylor & Francis Group
LONDON AND NEW YORK

First published 2020
by Routledge
2 Park Square, Milton Park, Abingdon, Oxon OX14 4RN

and by Routledge
52 Vanderbilt Avenue, New York, NY 10017

Routledge is an imprint of the Taylor & Francis Group, an informa business

British Library Cataloguing-in-Publication Data
A catalogue record for this book is available from the British Library

Library of Congress Cataloging-in-Publication Data
A catalog record for this book has been requested

ISBN: 978-0-367-28077-2 (hbk)
ISBN: 978-0-429-29955-1 (ebk)

Typeset in Times New Roman
by Apex CoVantage, LLC

Contents

Figures

Tables

About the authors

Andrew Chater (PhD, 2016, Western University) is an assistant professor (limited term) at Brescia University College. He is also a fellow at Polar Research and Policy Initiative and the 2019 Fulbright Visiting Research Chair in Arctic Studies at the University of Washington. He is researching Arctic governance and Canadian foreign policy in the Arctic. Recent publications have appeared in *The Canadian Journal of Communication*, *The Northern Review*, *International Organisation Research Journal*, and *Strategic Analysis*.

P. Whitney Lackenbauer (PhD, 2004, University of Calgary) is the Canada Research Chair (Tier 1) in the Study of the Canadian North at Trent University, Honorary Lieutenant-Colonel of 1st Canadian Ranger Patrol Group, co-director of the Centre on Foreign Policy and Federalism, and co-editor of the Documents on Canadian Arctic Sovereignty and Security (DCASS) series. His recent books include *China's Arctic Ambitions and What They Mean for Canada* (co-author 2018), *Canadian Armed Forces Arctic Operations, 1945–2015: Historical and Contemporary Lessons Learned* (co-editor 2017), and *Le Canada et l'Arctique* (co-author 2017).

Mathieu Landriault (PhD, 2013, University of Ottawa) is the director of the Observatoire de la Politique et la Sécurité de L'Arctique (OPSA), based in Montreal. He currently teaches at the School of Political Studies at the University of Ottawa, as well as at the School of Conflict Studies at Saint Paul University. He is also an associate researcher at the Center for Interuniversity Research on the International Relations of Canada and Quebec (CIRRICQ). He is researching Arctic security, sovereignty, and governance issues in the circumpolar region in general and the Canadian Arctic in particular, as well as Arctic paradiplomacy.

Elana Wilson Rowe (PhD, 2006, Cantab) is Research Professor at the Norwegian Institute of International Affairs (NUPI). She is also an adjunct professor at the High North Centre for Business and Governance at Nord University (Bodø). Wilson Rowe has authored articles on the politics of expertise in global governance, Russian foreign and northern policies, and climate change negotiations. She is the author of *Arctic Governance: Power in Cross-Border Co-operation* (2018), and *Russian Climate Politics: When Science Meets Politics* (2013).

Acknowledgments

Mathieu Landriault would like to acknowledge Paul Minard for his valuable insights, which helped stimulate reflection on key aspects of this book. Additionally, the author would like to sincerely thank Rachel Côté for her patient support and help given through multiple readings of this manuscript; this book would not have been possible without her devoted assistance.

Andrew Chater would like to thank: Elizabeth Riddell-Dixon, Adam Harmes, Gordon McBean, and Erika Simpson, for supporting this research in its early stages; Patricia Bell, Cindy Dickson, Terry Fenge, Chief Gary Harrison, Chief Ruth Massie, and Darlene Nuqingaq, for supporting northern travel crucial to this research; all of the interviewees, for giving generously of their time; and Al and Susan Chater, for endless support and inspiration.

P. Whitney Lackenbauer would like to thank the Canada Research Chair program for research funding, and Jennifer Arthur-Lackenbauer and their three sons for ongoing support and tolerance of his frequent travel. Ongoing conversations with Andrea Charron, Rob Huebert, Peter Kikkert, Rosemarie Kuptana, Suzanne Lalonde, Bridget Larocque, Heather Nicol, Andreas Østhagen, and many others (including the co-authors of this book) continue to challenge and (re)shape his ideas on sovereignty and Arctic governance.

Elana Wilson Rowe's contribution to the book on the private sector in Arctic governance was supported by the POLGOV (Science and Business in Arctic Governance) project, which was funded by the Norwegian Research Council (project no. 257664). She would like to thank members of the POLGOV team – Indra Øverland, Svein Vigeland Rottem, Christian Prip, Michael Denison, Emma Wilson, Martin Sommerkorn, and Ilan Kelman – for inspiring and insightful discussions on topics related to this book.

Finally, the authors would like to express gratitude to series editor Timothy Shaw, and to Routledge editor Robert Sorsby, for the invitation to publish our analysis of Arctic governance.

Acronyms

A5	Arctic 5
AAC	Arctic Athabaskan Council
AARI	Arctic and Antarctic Research Institute
AC	Arctic Council
ACA	Arctic Circle Assembly
ACAP	Arctic Contaminant Action Plan
ACIA	Arctic Climate Impact Assessment
AEC	Arctic Economic Council
AECO	Association of Arctic Expedition Cruise Operators
AEPS	Arctic Environmental Protection Strategy
AIA	Aleut International Association
AIP	Arctic Investment Protocol
AMAP	Arctic Monitoring and Assessment Programme
AMSA	Arctic Marine Shipping Assessment
APS	American Polar Society
BBNJ	Biodiversity beyond national jurisdiction
BEAC	Barents Euro-Arctic Council
BEAR	Barents Euro-Arctic Region
BP	British Petroleum
BRC	Barents Regional Council
CAO	Central Arctic Ocean
CAOF	Central Arctic Ocean Fisheries
CAFF	Conservation of Arctic Flora and Fauna
CBF	Circumpolar Business Forum
CCU	Circumpolar Conservation Union
CLCS	Commission on the Limits of the Continental Shelf
CSR	Corporate Social Responsibility
DC	District of Columbia
EU	European Union
EEZ	Exclusive Economic Zone
FiSCAO	Scientific Experts on Fish Stocks in the Central Arctic Ocean
GCI	Gwich'in Council International
GMO	Genetically modified organisms

IADC	International Association of Drilling Contractors
IASC	International Arctic Science Committee
ICC	Inuit Circumpolar Council
IMO	International Maritime Organization
INQ	Institut Nordique du Québec
IPCC	Intergovernmental Panel on Climate Change
IPF	International Polar Foundation
ISO	International Organization for Standardization
NF	Northern Forum
NEFCO	Nordic Environmental Finance Corporation
NGO	Non-governmental organization
NOAA	National Oceanic and Atmospheric Administration
NORA	Nordic Atlantic Cooperation
OECD	Organisation for Economic Co-operation and Development
OGA	Oil and Gas Assessment
OGP	Association of Oil and Gas Producers
PN	Plan Nord
PQ	Parti Québécois
RAIPON	Russian Association of Indigenous Peoples of the North
RFMO/A	Regional Fisheries Management Organization/Arrangement
SAO	Senior Arctic Officer
SC	Saami Council
SHE	Hot Spots Exclusion
SNP	Scottish National Party
TFOPP	Task Force on Arctic Marine Oil Pollution Prevention
TFTIA	Task Force on Telecommunications Infrastructure in the Arctic
UN	United Nations
UNCLOS	United Nations Convention on the Law of the Sea
UNESCO	United Nations Educational, Scientific and Cultural Organization
UNFCCC	United Nations Framework Convention on Climate Change
UNFSA	United Nations Fish Stocks Agreement
US	United States
WEF	World Economic Forum
WWF	World Wide Fund for Nature (formerly World Wildlife Fund)

Introduction

It has become cliché to start any study of the Arctic by describing it as a region undergoing profound transformation, and almost ritualistic for Arctic analysts to stress the radical changes brought about by climate change and global warming. These global phenomena are reshaping the Arctic's physical environment, and simultaneously changing its economic, social, cultural, and political settings.

The melting of the ice cover has opened the door for industry to pursue new opportunities, and for shipping (and naval) interests to consider Arctic routes as shortcuts for international navigation. At the same time, these activities run the risk of disarticulating cultural practices of Northerners – particularly Indigenous peoples – and of undermining food security and traditional food supplies.

A diverse set of governance mechanisms at multiple levels politically manage the changing environment. The two voyages of the luxury cruise ship *Crystal Serenity* through the Canadian Arctic in the summers of 2016 and 2017 serve as a case in point. Crystal Cruise, the parent company, consulted with local communities in planning its itinerary, agreed to some restrictions on the use of heavy fuel, and charted a course that avoided sensitive ecosystems (Comer *et al.*, 2016; McWhinnie *et al.*, 2018). The Canadian federal government closely monitored the voyages, and relations between the Government of Canada and the company were of a co-operative nature. The territory of Nunavut, a sub-national jurisdiction, also cleared the transits, although the first voyage stirred a review of regulations of mass tourism in the territory (Government of Nunavut, 2016). Ultimately, the company decided to suspend further activities in the Canadian Arctic in order to build the first cruise ship that would comply with the Polar Code – a list of requirements that the International Maritime Organization agreed upon in 2014. The *Crystal Serenity* voyages illustrate stakeholder multiplicity in the Arctic region: a national government, a sub-national government, local communities, a corporation, and an international organization all played a role in the elaboration and implementation of the cruise ship's transit of Canada's Arctic waters.

Despite the complexity of Arctic governance, existing scholarship predominantly focuses on the politics of and relations between states. We have a strong baseline on the foreign policy preferences of the Arctic states themselves (e.g. Griffiths *et al.*, 2011 on Canada; Jensen, 2013 on Norway; Laruelle, 2013, and Sergunin and Konyshev, 2015 on Russia; Wilson Rowe, 2009, 2017). To move towards thinking systematically about Arctic governance, the standard approach amalgamates the Arctic states' circumpolar interests (or find the points of disagreement) while assessing their past, present, and future policies and decisions towards the region. Scholars have worked to expand this analysis by accounting for the development of different regimes and policy framings (Hoel, 2015; Hønneland and Stokke, 2010; Powell and Dodds, 2014; Steinberg *et al.*, 2015), by analyzing Arctic politics as compared to broader political trends in global governance (Dodds and Nuttal, 2016; Tennberg, 2015; Wilson Rowe, 2018), and by holding the analysis up against different schools of thought in international relations (Byers, 2018). Increasing attention is paid to specific groups in Arctic politics, particularly the roles of Indigenous peoples' organizations, non-state actors, and non-Arctic states in shaping Arctic governance (Knecht, 2016; Shadian, 2014; Spence, 2016; Wehrmann, 2017), although there has been little cross-group comparison. A recent deluge of studies seek to discern the Arctic interests of non-Arctic states, especially Asian and European states (Jakobson and Melvin, 2016; Lackenbauer *et al.*, 2018; Murray and Nuttall, 2014; Sinha and Bekkevold, 2015). The Arctic Council (AC), as a regional political forum, garnered significant attention and generated extensive scholarship, covering state diplomacy as well as other group interests and interventions (Chater and Landriault, 2016; Dodds, 2013; English, 2013; Gracyzk and Koivurova, 2014; Nicol and Heininen, 2013; Pedersen, 2012; Rottem, 2013).

This book argues that confining our understandings of Arctic governance to Arctic states and a focus on the AC as the primary site of circumpolar governance perpetuates an incomplete picture. Instead, this book embraces the complexity of governance in the Arctic by systematically analyzing and comparing the position, interventions, and influence of different actor groups seeking to shape Arctic political and economic outcomes in multiple sites of Arctic politics (formal and informal). Businesses, sub-national governments, non-governmental organizations, Indigenous peoples, international organizations, and scientists are stakeholders (or rightsholders) as well as active actors. These actors interact with one another and the state in various settings. The chapters that follow focus on Arctic circumpolar and international politics, while leaving the interplay of these groups at the national/domestic and community levels to further analysis. In the concluding chapter, we focus on the interplay between the different actor groups analyzed in each chapter, exploring how they interact with one another in a broader quest for authority to define Arctic challenges and prescribe solutions.

States and "other" actors

The predominant role of the state needs to be understood to make sense of the roles occupied by other actors, as well as their influence. A state-centric approach makes sense when considering the dominant role played by sovereign states in the economic and social development of the region, given that state guidance and assistance often remains pivotal to development projects in sparsely populated and remote, austere operating environments. Moreover, Arctic coastal states have yet to sort out areas of overlap on extended continental shelves in the Arctic basin. Accordingly, media commentators have described the region as a potential site of state conflict and competition, awakening realist geopolitical warnings about instability and power dynamics. This narrative of risk remains dominant, even though it is frequently opposed by Arctic states that trumpet the robust governance regime promoting and sustaining regional peace and development (Steinberg *et al.*, 2015; Wegge, 2012).

From this departure point in narratives of instability and risk, it is logical to assert that "Arctic states remain the dominant actors in Arctic governance" and that "sovereign rights are at the centre of discussions on the future of the Arctic" (Melvin and Bergh, 2016: 4–5). We contend that this assessment overstates sovereignty concerns and misses a crucial turn in regional dynamics, wherein ownership questions tend to be sidelined for pragmatic discussions about regulation and management. Consequently, from some perspectives and in certain policy fields, the regional political landscape has evolved from a sovereignty-focused to a governance-concerned one.

This transition to broader governance processes has increased the space available for a broad range of actors to shape key domains of Arctic life and politics. Drawing on developments in the broader literature on global governance, we do not consider states and other actors to be in competition or engaged in a zero-sum game over influence. In other words, the influence exercised by non-state actors does not come inherently at the expense of the state and does not result in the "retreat of the state." Indeed, the influence of non-state actors can sometimes be highly symbiotic with or reliant upon relations to states (Neumann and Sending, 2010). A second caveat relates to this idea of authority. Earlier research in global governance posited that different "kinds" of actors possessed different forms of authority. For example, non-governmental organizations (NGOs) or Indigenous peoples' organizations, representing affected parties and the original peoples of the Arctic, exercised "moral authority" based on their status, whereas scientists exercised "expert authority" and international organizations a "delegated authority" derived from their membership (Barnett and Finnemore, 2004). The chapters that follow examine instead the influence of different actor groups without positing at the outset that non-state actors engage in regional and global governances in fundamentally different ways than state actors. Most chapters pose relatively open questions about the positioning of different actor groups and examine how and why their positions or resources are used in specific contexts to influence Arctic governance.

Defining governance

Conceptualized as a co-operative patchwork, scholars regularly invoke the concept of governance to describe Arctic co-operation. The term is invariably present in most scholarly endeavours focusing on rules, norms, and political decisions in the circumpolar world. Its meanings and forms, however, are usually taken for granted or rarely explicitly conceptualized.

At its core, the concept of governance is about regulating and bringing order to human collectivities (Scholte, 2008). Invariably, rules are at the center of what we mean by governance. Following Roger and Dauvergne (2016), we define governance as the processes by which rules, standards, and regulations are crafted, adopted, and enforced by political, economic, or social actors (416).

The study of governance processes necessarily considers a multiplicity of actors interacting and influencing the policy outcome. Ruhanen *et al.* (2010) emphasize that governance involves many stakeholders, leading to heightened unpredictability and uncertainty (5–6). Hence, governance by its very nature involves a complex web of interrelations between different types of actors contributing to the rule-making process. Some of these actors are more powerful or influential than others, but no single actor can completely impose its will on all others, even though a state or a group of states can play a substantial coordination or oversight role in the process. As Jönsson (2008) observes, the notion of governance conceptualizes that "order is based neither on anarchy, nor on hierarchy" (30). Instead, we are confronted with co-operation and competition patterns that do not fit into simplistic power relations. Accordingly, governance is interested in the nature of the decisions taken and the rules adopted, as well as the nature of the process that generates these norms and rules. Thus, questions about accountability, transparency, and the involvement of interested parties are legitimate grounds of exploration in governance studies (Ruhanen *et al.*, 2010).

Diverse stakeholders can play different roles in governance processes. The most visible function is regulatory, in which actors can define regulations, permissions, and prohibitions and enforce them. State actors and international organizations are typically concerned with this type of prerogative, but governance actors can assume other important roles as well. For example, Young (2002) points out that stakeholders can act as catalysts by bringing interested parties together, facilitating co-operation and exchanging views. In other cases, both state and non-state actors can fund or facilitate the creation of new knowledge and understandings about specific issues or a region. Non-state actors, including corporations, civil society, and academics, have often performed this generative role. Finally, stakeholders can contribute in a programmatic fashion by pooling resources to take concrete action in a given policy area.

Scholars have conceptualized various types of governance. Intergovernmental governance refers to processes in which only governments are involved. Transnational governance is defined as processes in which non-state actors adopt rules that seek to move behavior towards a shared, public goal in at least two states (Roger and Dauvergne, 2016: 416). (It is important to note that non-state actors can play a significant role in intergovernmental governance, while state actors can influence non-state actors in transnational governance processes.) Private governance is when "the interactions among private actors, or between private actors on the one hand and civil society and state actors on the other, give rise to institutional arrangements that structure and direct actors' behavior in an issue-specific area" (Falkner, 2003: 72). The creation of certifications under the International Organization for Standardization (ISO) is a clear example of corporate governance practice (see Chapter 5). We add that private governance is comprised of rules and norms adopted internally by corporations when dealing with issues such as accountability, or consultations with local communities.

Finding Arctic governance: from global to local governance

Although circumpolar region building has been going on for decades, the Arctic as a region remains a project in construction, particularly at the institutional level. Rules and standards can be discerned at different levels of governance, leaving the region a highly fragmented governance space (Young, 2016).

Global governance is a cornerstone of the Arctic governance infrastructure, represented by two types. First, global agreements are concerned with the entire planet, including the Arctic region. In this respect, the region is inextricably implicated in global dynamics and subjected to global campaigns and agreements. In a region comprised of states surrounded by an ocean, the United Nations Convention on the Law of the Sea (UNCLOS), signed in 1982, is central to Arctic governance. UNCLOS, as the "constitution of the sea," codifies states' rights and obligations on a myriad of issues, from international navigation and transit rights to the delimitation of territorial waters and continental shelves. Arctic states recognize the legitimacy of these rules, and the nature of their application tends to dominate state discussions on Arctic shipping, resource exploitation, and territorial delimitation. Other global agreements include the United Nations Convention on Biological Diversity, the International Convention for the Prevention of Pollution from Ships, the Convention on Migratory Species of Wild Animals, and the United Nations Fish Stocks Agreement. These codify rules worldwide, including the Arctic. Although these global agreements can be tailored to address regional specificities, the law must be applied in a manner consistent with the legal intent of the original convention.

Second, international organizations have developed agreements specific to the Arctic region. In this form of governance, global institutions and their extensive membership act as convenors and craft region-specific agreements. We might consider this an intermediate form of governance between global and regional, in which global institutions generate rules and norms applicable in specific regional settings. This type of governance includes efforts by organizations with global outreach (such as the ISO for companies) or sectoral corporate associations (such as the Cruise Lines International Association) to elaborate Arctic-specific codes of conduct and standards.

Local governance occupies the other side of this continuum. Far from global rules and standards, local governance processes involve stakeholders at the community level to tackle problems facing people in their day-to-day lives. The mechanisms by which people in specific areas make or shape the rules vary across the Arctic region, with counties in Norway holding different legal rights and powers than Indigenous land claim settlement regions in Alaska or Northern Canada. For analytical purposes, local governance does not involve coordination of efforts between communities in a given region, with this concerted effort falling into the category of regional governance. For example, the governance exercised by communities in Nunavut, Canada, is appropriately situated as local governance, whereas policy recommendations issued by the Inuit Circumpolar Council (ICC) constitute an effort at developing rules at the regional level.

Regional governance, for its part, seeks to coordinate local efforts while developing regional rules and standards. As such, Arctic institutions face pressures on two fronts. On the one hand, they must be inclusive and incorporate local stakeholders in the rule-making process, at the risk of slowing down governance processes and encountering low levels of consensus given divergent local realities. Nonetheless, inclusiveness can counter accusations that regional governance practices are disconnected from local inhabitants, given that legitimacy relies on the credibility of a regional actor or mechanism in representing or reflecting local concerns. On the other hand, regional governance structures must carve a position for themselves in relation to global governance institutions by stressing the particularities of the region, thus establishing why a regional solution (rather than a global one) is necessary to address regional problems. As we emphasize in Chapter 2, Arctic states – and particularly the "Arctic 5" of Canada, the United States, Russia, Denmark, and Norway – are adamant that they be respected as preponderant actors in the region by "external" actors (whether European or Asian states) seeking to be part of the governance exchange. Differing interpretations or definitions of who is habilitated to speak in regional governance processes (all Arctic states? only coastal Arctic states? user states?) may trigger competition amongst or confrontation between rightsholders and other stakeholders.

Of course, regional actors still consider global governance agreements to be relevant. In a region underdeveloped at the institutional level and overlapping with other regions (Europe, Scandinavia, and North America), UNCLOS and global conventions provide a set of clear and solid norms to guide regional interactions. Nevertheless, the unique Arctic environment has necessitated for distinct regional governance processes. The AC, established through the Ottawa Declaration in 1996, is the most developed institution that coordinates regional governance efforts (see Chapter 1). Other forums, such as the Arctic Economic Council (see Chapter 5), the Arctic Circle Assembly (see Chapter 1), and the Arctic Frontiers meetings, mobilize regional stakeholders to promote rules and norms. The ICC is an example of regional governance emanating from civil society (see Chapter 4), while the Association of Arctic Expedition Cruise Operators is one with corporate origin (see Chapter 5). Scientists hold unique power in the region, as represented in several institutions.

Structure of this book

This book focuses on emerging forms of governance in the Arctic region, exploring how different types of state and non-state actors promote and support rules and standards. For clarity and conciseness, we analyze neither governance processes at the local level nor debates taking place primarily at a national level on Arctic issues. Rather, we concentrate on governance processes taking place at a circumpolar transnational level, involving actors from multiple states.

In Chapter 1, we focus on regional governance initiatives put forward by Arctic states. The evolution of the Arctic Council, but also of new emerging forums such as the Arctic Circle Assembly and the Arctic Economic Council, will be front and center.

In Chapter 2, we study the creation and evolution of the Arctic 5 ensemble and the Arctic 5+5 as a new form of intergovernmental governance. The moratorium on commercial fisheries in the Central Arctic Ocean acts as a case study to anticipate future prospects for intergovernmental co-operation outside the Arctic Council.

Chapter 3 explores intergovernmental governance of a different kind. Sub-national governments engage in paradiplomacy to further their interests and to develop ties with like-minded entities, in order to develop a coherent international presence. As a result, the opportunities and limitations offered to sub-national governments will be scrutinized, for both long-standing Arctic players and institutions (Lapland, Russian states, Alaska, Northern Forum) and new emerging ones (Québec, Scotland).

Chapter 4 analyzes transnational governance and the role of non-state organizations (NGOs) in Arctic governance, focusing on indigenous peoples' organizations and civil society groups/NGOs. The Arctic region has experienced a high level of civil society inclusion in the main governance institutions. It is therefore crucial to assess civil society's contribution and efforts to develop comprehensive rules to influence both state and private actor behaviour.

Chapter 5 focuses on private governance. Companies and associations of private interests are now active players, each developing internal rules and norms to guide their Arctic activities and participating in multilateral forums. In many cases, these companies will be at the forefront of future economic developments in or affecting the region.

The book concludes with an evaluation of the future of Arctic governance, considering the multiplicity and diversity of actors involved. The great diversity of actors highlights the necessity to analyze the specific contributions of and roles played by every type of actor. Arctic states are key actors in regional governance, but they are not alone.

Note on methods

We employ two main methods in this book: process tracing and descriptive statistical analysis. Process tracing involves constructing key sequences of events from primary sources and inferring causality in these cases. Examples include recounting the sequence of events at international negotiations, or describing the underlying motivations for a particular action in an international negotiation, drawing on records of those meetings. Descriptive statistical analysis includes inferring tendencies from data describing key trends and events. Examples include reconstructing the composition of meeting participation or recapping key economic indicators.

Data comes from primary documents freely available online from sources such as the websites of the Arctic Circle Assembly, Arctic Council, Arctic Economic Council, Arctic Frontiers, Arcticnet, Barents-Europe Arctic Co-operation, International Maritime Organization, and the Northern Forum. Examples of documents include meeting minutes, agendas, policy briefs, reports, memorandums, declarations, and participant lists. Additional insights come from secondary sources, such as news articles and the work of other scholars. Andrew Chater contributed unpublished insights from 33 semi-structured interviews that he conducted in 2013 with key Arctic Council policy-makers (government officials and diplomats, Indigenous peoples' organizations, and consultants) for a different project entitled *Explaining the Evolution of the Arctic Council*. The interviews took place in person in Ottawa and Washington, DC, as well as by phone and email.

References

Barnett, Michael N. and Martha Finnemore. 2004. *Rules for the World: International Organizations in Global Politics*. Ithaca, NY and London: Cornell University Press.

Byers, Michael. 2018. "Crises and International Co-operation: An Arctic Case Study." *International Relations* 31, no. 4: 375–402.

Chater, Andrew and Mathieu Landriault. 2016. "Understanding Media Perceptions of the Arctic Council." *Arctic Yearbook* 5: 60–72.

Comer, Bryan, Naya Olmer, and Xiaoli Mao. 2016. "Heavy Fuel Oil Use in Arctic Shipping in 2015." *Working Paper 21*. Washington, DC: The International Council on Clean Transportation. www.theicct.org/sites/default/files/publications/HFO%20Arctic%20Shipping_working-paper_vF_21102016.pdf.

Dodds, Klaus. 2013. "Anticipating the Arctic and the Arctic Council: Pre-emption, Precaution and Preparedness." *Polar Record* 49, no. 2: 193–203.

Dodds, Klaus and Mark Nuttall. 2016. *The Scramble for the Poles: The Geopolitics of the Arctic and the Antarctic*. London: Polity Press.

English, John. 2013. *Ice and Water: Politics, People and the Arctic Council*. Toronto, Ontario: Penguin Random House Canada.

Falkner, Robert. 2003. "Private Environmental Governance and International Relations: Exploring the Links." *Global Environmental Politics* 3, no. 2: 72–87.

Government of Nunavut. 2016. "Responsible Cruise Tourism Development in Nunavut." *Northern Public Affairs*, October. www.northernpublicaffairs.ca/index/responsible-cruise-tourism-development-in-nunavut/.

Gracyzk, Piotr and Timo Koivurova. 2014. "A New Era in the Arctic Council's External Relations? Broader Consequences of the Nuuk Observer Rules of Governance." *Polar Record* 50, no. 3: 225–236.

Griffiths, Franklyn, Rob Huebert, and P. Whitney Lackenbauer. 2011. *Canada and the Changing Arctic: Sovereignty, Security and Stewardship*. Waterloo, Ontario: Wilfrid Laurier University Press.

Hoel, Alf Håkon. 2015. "Oceans Governance, the Arctic Council and Ecosystem-Based Management." In *Handbook of the Politics of the Arctic*, edited by Leif Christian Jensen and Geir Hønneland, 265–280. Cheltenham, UK: Edward Elgar Publishing.

Hønneland, Geir and Olav Schram Stokke (editors). 2007/2010. *International Co-operation and Arctic Governance: Regime Effectiveness and Northern Region Building*. London: Routledge.

Jakobson, Linda and Neil Melvin (editors). 2016. *The New Arctic Governance*. Oxford: Oxford University Press.

Jensen, Leif C. 2013. "Seduced and Surrounded by Security: A Post-Structuralist Take on Norwegian High North Securitizing Discourses." *Cooperation and Conflict* 48, no. 1: 80–99.

Jönsson, Christer. 2008. "Global Governance: Challenges to Diplomatic Communication, Representation, and Recognition." In *Governance and Diplomacy: Worlds Apart?*, edited by Andrew F. Cooper, William Maley, and Brian Hocking, 29–38. New York: Palgrave Macmillan.

Knecht, Sebastian. 2016. "The Politics of Arctic International Co-operation: Introducing a Dataset on Stakeholder Participation in Arctic Council Meetings, 1998–2015." *Co-operation and Conflict* 52, no. 2: 203–223.

Lackenbauer, P. Whitney, Adam Lajeunesse, James Manicom, and Frédéric Lasserre. 2018. *China's Arctic Ambitions and What They Mean for Canada*. Calgary, Alberta: University of Calgary Press.

Laruelle, Marlene. 2013. *Russia's Arctic Strategies and the Future of the Far North*. Armonk, NY: M.E. Sharpe.

McWhinnie, Lauren, William Halliday, Stephen Insley, Casey Hilliard, and Rosaline Canessa. 2018. "Vessel Traffic in the Canadian Arctic: Management Solutions for Minimizing Impacts on Whales in a Changing Northern Region." *Ocean and Coastal Management* 160: 1–17.

Melvin, Neil and Kristopher Bergh. 2016. "Introduction." In *The New Arctic Governance*, edited by Neil Melvin and Kristopher Bergh, 1–12. Oxford: Oxford University Press.

Murray, Robert and Anita Dey Nuttal (editors). 2014. *International Relations and the Arctic: Understanding Policy and Governance*. Amherst, NY: Cambria Press.

Neumann, Iver B. and Ole Jacob Sending. 2010. *Governing the Global Polity*. Ann Arbor, MI: University of Michigan Press.

Nicol, Heather and Lassi Heininen. 2013. "Human Security, the Arctic Council and Climate Change: Competition or Co-existence?" *Polar Record* 50, no. 1: 80–85.

Pedersen, Torbjørn. 2012. "Debates Over the Role of the Arctic Council." *Ocean Development and International Law* 43, no. 2: 146–156.

Powell, Richard C. and Klaus Dodds (editors). 2014. *Polar Geopolitics: Knowledges, Resources and Legal Regimes*. Cheltenham, UK: Edward Elgar Publishing.

Roger, Charles and Peter Dauvergne. 2016. "The Rise of Transnational Governance as a Field of Study." *International Studies Review* 18, no. 3: 415–437.

Rottem, Svein Vigeland. 2013. "The Arctic Council and the Search and Rescue Agreement: The Case of Norway." *Polar Record* 50, no. 3: 284–292.

Ruhanen, Lisa, Noel Scott, Brent Ritchie and Aaron Tkaczynski. 2010. "Governance: A Review and Synthesis of the Literature." *Tourism Review* 65, no. 4: 4–16.

Scholte, Jan Aart. 2008. "From Government to Governance: Transition to a New Diplomacy." In *Governance and Diplomacy: Worlds Apart?*, edited by Andrew F. Cooper, William Maley, and Brian Hocking, 39–60. New York: Palgrave Macmillan.

Sergunin, Alexander and Valery Konyshev. 2015. *Russia in the Arctic: Hard or Soft Power*? Stuttgart, Germany: Ibidem Press.

Shadian, Jessica M. 2014. *The Politics of Arctic Sovereignty: Oil, Ice and Inuit Governance*. Abingdon, UK: Routledge.

Sinha, Uttam Kumar and Jo Inge Bekkevold. 2015. *Arctic: Commerce, Governance and Policy*. New York: Routledge.

Spence, Jennifer. 2016. Finding a Place in the Arctic Council for Non-Arctic Actors: A Social Network Analysis of the Arctic Monitoring and Assessment Programme. *Arctic Yearbook* 5: 74–90.

Steinberg, Philip E., Jeremy Tasch, and Hannes Gerhardt. 2015. *Contesting the Arctic: Politics and Imaginaries in the Circumpolar North*. London: I. B. Tauris.

Tennberg, Monica. 2015. "Arctic Change Through a Political Reading." In *Handbook of the Politics of the Arctic*, edited by Leif Christian Jensen and Geir Hønneland, 408–420. Cheltenham, UK: Edward Elgar Publishing.

Wegge, Njord. 2012. "The EU and the Arctic: European Foreign Policy in the Making." *Arctic Review* 3, no. 1: 6–29.

Wehrmann, Dorothea. 2017. "Non-State Actors in Arctic Council Governance." In *Governing Arctic Change: Global Perspectives*, edited by Kathrin Keil and Sebastian Knecht, 187–206. Basingstoke, UK: Palgrave MacMillan.

Wilson Rowe, Elana (editor). 2009. *Russia and the North.* Ottawa, Ontario: University of Ottawa Press.

Wilson Rowe, Elana. 2017. "The Arctic in Moscow." In *Sustaining Russia's Arctic Cities: Resource Politics, Migration and Climate Change*, edited by Robert W. Orttung, 25–41. New York: Berghahn Books.

Wilson Rowe, Elana. 2018. *Arctic Governance: Power in Cross-Border Relations*. Manchester, UK: Manchester University Press.

Young, Oran. 2002. "Can the Arctic Council and the Northern Forum Find Common Ground?" *Polar Record* 38, no. 207: 289–296.

Young, Oran. 2016. "Governing the Arctic Ocean." *Marine Policy* 72: 271–277.

1 Arctic states and regional governance

Arctic states have been pivotal actors in regional governance, representing the main architects of the existing institutional infrastructure. Multilateralism and pragmatic co-operation between these states have dominated regional dynamics. At the same time, Arctic states have been adamant about promoting their predominant role and stature, especially in relation to non-Arctic states. We will see that this posture has evolved in recent years, leading to institutional reforms adding new pieces to Arctic governance.

As the amount of ice in the Arctic has diminished, the diversity of governance institutions for the Arctic region has increased concurrently. In 1996, the Arctic Council (AC) was essentially a governmental environmental research forum that issued occasional policy recommendations. Today, it also serves as a broader policy-making body for the Arctic region. In 1996, environmental issues such as persistent organic pollutants and contaminants dominated policy conversations. Today, climate change has multiplied the environmental issues with which the AC grapples, as well as the economic opportunities in the region. In 1996, the AC stood as the main venue for conversation on the transnational governance of the Arctic, akin to an Arctic United Nations dominated by its eight members: the Arctic states (Canada, Denmark, Finland, Iceland, Norway, Russia, Sweden, and the United States). Today, new sites for conversation, such as the Arctic Circle Assembly (ACA) and the Arctic Economic Council (AEC), have emerged, providing new platforms and forums for a broad array of academic, industry, regional, and non-Arctic actors to contribute to solving key regional issues.

How have the regional governance initiatives put forward by the Arctic sovereign states changed over time? Focusing on the AC, the ACA, and the AEC, this chapter reveals how the AC evolved over the last two decades, and why the ACA and AEC emerged to complement it.

Arctic ultilateralism – the Arctic Council

The Arctic Council represents the primary body for regional governance in the circumpolar north. Four key observations characterize the role and evolution of the AC in the broader Arctic governance framework as it has evolved over the last quarter century. First, the AC began as a limited environmental governance body created and dominated by the Arctic states. Second, there were early disagreements about its role in policy-making and the elasticity of its mandate. Third, the AC's policy-making role and mandate have grown, assuming not only generative and catalyst functions but also a regulatory role in harmonizing national policies. While the role changed, however, the international character of the body did not. Fourth, new forums have emerged that provide space for a broad array of actors to contribute to Arctic governance discussions and, some commentators believe, challenge the pre-eminence of the AC.

Early assessments of the AC observed a weak and uncertain body. Evan Bloom, an American diplomat involved in the council's creation, described the AC in 1999 as:

> A forum without legal personality, and thus not as an 'international organization' as that term is understood under international law, was an objective of the United States and is consistent with a tendency in recent American diplomatic practice to seek an informal co-operative structure when that structure is adequate for the purposes of the issues involved.
>
> (721)

The year before, political scientist Huebert (1998) questioned "whether or not this effort can be sustained, and whether Canada will maintain its leadership role" (57). Legal scholars Koivurova and Vanderzwaag (2006) were pessimistic, highlighting that in the AC's work "a 'study and talk' mentality has prevailed with numerous projects and workshops promoted and carried out" (191). Young (2005) offered more optimism, emphasizing that "the momentum associated with the growth of regional co-operation in the Arctic is sufficiently strong and the emergence of an Arctic-specific agenda is sufficiently clear to provide some basis for optimism regarding the success of region building in the region" (14).

Various scholarly works have called for an expanded AC with a mandate broadened beyond its historic service as an environmental research forum. Byers (2012) puts forward 11 recommendations, including the creation of an oil spill prevention treaty, a black carbon treaty, and a "negotiation of a treaty to demilitarize the surface of the Central Arctic Ocean" (23). The Munk-Gordon Arctic Security Program, a policy institute associated with the University of Toronto, echoes Byers's call and advocates that the AC mandate grow to include military security. While we do not share this view that expanding the AC's mandate will make it stronger, we argue that it is necessary to look beyond the Council to understand the full breadth of Arctic governance bodies and embrace new non-statist institutions.

The AC began as a limited environmental governance body created and dominated by the Arctic states, emerging out of a genuine desire among the Arctic states to address regional environmental issues and to ease Cold War tensions. The superpower struggle between the Soviet Union and the United States defined international conflict throughout the second half of the 20th century. In response, some actors embraced Arctic collaboration as a mechanism to help defuse the situation. As such, the AC emerged from a late-twentieth-century international context of a growing global environmental consciousness, thawing Cold War relations, and a belief that the Arctic offered a venue to defuse tensions.

In 1983, the United States government unveiled an Arctic policy that opened the door to collaboration with the Soviet Union in the region, citing enthusiasm for "mutually beneficial international co-operation in the Arctic" ("United States Arctic Policy"). Two years later, political scientist Oran Young called on Arctic governments to declare the region as a "zone of peace" (174). In October 1987, following an agreement on the reduction of nuclear stockpiles, Russian Premier Mikhail Gorbachev gave a speech in Murmansk, Russia that proposed designating the Arctic region a "zone of peace" to follow co-operation on environmental protection (Gordon Foundation, 2010: 1).

The Murmansk speech represented a clear breakthrough. Gorbachev affirmed genuine interest in environmental protection: the 1986 nuclear disaster at Chernobyl, a meltdown that resulted in hundreds of deaths and the abandonment of a town of nearly 50,000, raised the fear that Russia's lax environmental standards would result in another transboundary catastrophe (English, 2013: 10). Still, officials in various governments doubted the seriousness of Gorbachev's "zone of peace" proposal. If truly earnest, the Russian Premier might have proposed the closure of Murmansk's large military base. Nevertheless, the government of Finland, a former colony of Russia, sought to use the opportunity to improve relations between its allies and former colonizer. Officials pushed for a nuclear submarine agreement (in 1983), held a peace summit (in 1987), and finally hosted a conference on Gorbachev's idea in Rovaniemi, Finland in September 1989. This meeting led to the creation of the Arctic Environmental Protection Strategy (AEPS) in June 1991 (Young, 1998; Keskitalo, 2004; English, 2013), through which state scientists met bi-annually to discuss co-operation on environmental issues. The following five years saw the co-creation and exchange of new knowledge on items such as polychlorinated biphenyls (PCBs) and the life expectancy of Arctic residents. The success of the AEPS naturally led to discussions to elevate it from a strategy to an institution.

Negotiations began in 1995 to create a limited environmental governance body. Although all of the Arctic countries shared a vested interest in improving the state of the environment, Russia had the weakest environmental regulations. Given the Nordic countries' close proximity to Russia, they strongly advocated for a Council (interviews 8 and 7) that would help keep Russia accountable (interview 10). Canadian officials agreed with the Nordic governments, and were optimistic that the AC could be a forum to give a voice to Arctic residents. Accordingly, Canadian and Nordic leaders sought a strong AC that would include representatives of Arctic Indigenous peoples as full members and that would produce binding decisions. The United States and Russia wanted a weaker, more *ad-hoc* body that, essentially, perpetuated the AEPS under a new name (interview 15).

During the negotiations, Russian officials seemed "extremely nonchalant" about environmental issues, espousing the view that "Russia is a gigantic country" (interview 7). In the midst of a recession which saw Russia's economy shrink by half in the early 1990s, Russia's government closed half of its environmental monitoring stations (interview 18) and downgraded its Ministry of the Environment from a ministry to a "state committee" (interview 10). While environmentalists (who tended to promote democracy in post-Soviet Russia) remained bitter that the Soviet system had failed to protect the environment and to prevent the Chernobyl disaster, "major oligarchs and industrial concerns" opposed environmental regulation and democracy (interview 10). Nevertheless, Russian officials acknowledged that participation in an AC would provide Russia with an opportunity to join the world community and secure international funding for environmental projects (interview 16).

A persistent strain of American thought viewed multilateral institutions and agreements with suspicion; US officials worried that a strong AC would impose undue restrictions on sovereignty and would become a venue for other states to criticize the United States (interview 1). The Inuit Circumpolar Council (ICC), for example, stridently opposed various US policies, such as the 1972 Marine Mammals Protection Act, which banned the import of seal products; and the Alaska Native Claims Settlement Act, which unilaterally settled outstanding Indigenous land claims in that state (interview 1). American officials feared that including Indigenous peoples as voting members in an AC, with similar status to the Arctic states, would create a precedent to do so in other forums and affect ongoing negotiations to produce a *United Nations Declaration on the Rights of Indigenous Peoples* (interview 14). Other environmental institutions already existed, and influential US officials believed that environmental issues required a global rather than a regional approach (interview 1). Nonetheless, the United States saw benefits in information sharing through an AC (interview 9).

Given this backdrop of state interests, the AC's creation as a limited, soft-law body, rather than a formal multilateral organization, should come as little surprise. The 1998 *Iqaluit Declaration* specified that its role was to "provide a means for promoting co-operation, coordination and interaction among the Arctic States . . . with the involvement of the Arctic Indigenous communities and other Arctic inhabitants on common Arctic issues, in particular issues of sustainable development and environmental protection in the Arctic," but not military security, as stated in a footnote (Arctic Council, 1998). Framers deliberately excluded traditional military security from the Council's mandate, but included sustainable development, thus adding a human dimension to the environmental focus of the AEPS. Environmental security became the AC's main focus, but the Arctic states' foreign affairs departments would administer the institution's work. Although the AC was not conceived as a formal policy-making body, nothing in its founding documents forbid the future possibility. It would not have a permanent secretariat, but the chairmanship of the institution would rotate every two years. Observer states and non-governmental organizations would attend AC meetings without a formal role or weight in deliberations. Indigenous peoples' organizations would participate as Permanent Participants at the main table as non-voting members – lacking the status of the Arctic states, but an international breakthrough nevertheless. As a consensus-based forum, the AC would hold bi-annual meetings to oversee progress on projects and share findings.

Four major takeaways about the founding of the AC emerge:

1. the AC is the major governance institution for the Arctic region because it includes all of the Arctic states as members;
2. it is not an international forum, but rather a regional forum, because the role of non-Arctic states is limited, largely absent in the founding of the institution;
3. the Arctic states dictate the structure and content of the institution; and,
4. the AC began life as an environmental body, but the reasons for the creation of the institution were not completely environmental in nature.

In the years that followed, the AC assumed a generative role by completing monitoring projects, drafting action plans, and producing science-based assessments, such as projects on biodiversity and persistent organic pollutants.[1] Disagreements arose regarding its role in policy-making and the elasticity of its mandate as states wrestled with the forum's limitations. At the AC's November 1999 meeting in Washington, the United States proposed to give a joint presentation with Russia on the Arctic Military Environmental Co-operation Group, which aimed to reduce the environmental impact of military vehicles in the Arctic. Canada's delegation vetoed the discussion immediately as falling outside the mandate of the AC because it concerned military security. At the same meeting, the delegations of Sweden and Norway proposed the creation of a formal policy to codify national commitments to mitigate PCB contamination in the *Arctic Contaminant Action Plan* (ACAP). The meeting minutes reported "that delegates should be prepared to negotiate final text for the strategy component of the [ACAP] plan at the next meeting" (with Finland, Iceland, and Denmark in support). Canada again vetoed the creation of formal policy as an intrusion on state sovereignty. "As far as using the ACAP as a coordinated approach in international fora, Canada stated that the Council must continue to rely on individual states to take action," the minutes noted. In support, the ICC asserted that the ACAP plan represented "a concerted effort to look after needs of indigenous people in the Arctic" (Arctic Council, 1999). These disagreements about the AC's role would continue.

Although the policy-making role and mandate of the AC has grown over the past two decades, the legal character of the body has not. The Council acts as a catalyst, assuming the role of convenor facilitating co-operation between member states and offering a co-operative platform to launch substantive initiatives. Global interests in the Arctic peaked in 2007, with actions by some member states spurring concerns about the prospect of heightened regional tensions. The Russian *Arktika* expedition's planting of a Russian flag at the bottom of the sea under the North Pole (see Chapter 2) attracted global attention, prompting other Arctic states to declare their intentions to improve their Arctic capabilities. Canadian Conservative Prime Minister Stephen Harper adopted a proactive stance which vigorously asserted Arctic sovereignty and announced more military investments in the region (Blunden, 2009; Lackenbauer, 2009; Dodds, 2011; Landriault, 2016).

Calls by non-Arctic actors, such as the European Union, for an Arctic treaty to ensure greater regional stability met resistance from the Arctic states

(Koivurova, 2008; Koivurova, 2010; Wegge, 2012). Instead, the Arctic states sought to inject more dynamism and ambition into existing regional governance institutions; regulatory and harmonization initiatives were now necessary. The creation of the Arctic 5 (A5) sub-ensemble and the A5's *Ilulissat Declaration* in 2008 were the results of this contestation. Arctic coastal states decided to reaffirm their predominant role in the region while reassuring global players by sending positive signals about the Arctic geopolitical situation (see Chapter 2).

The Nuuk Ministerial in 2011 marked a turning point in the history of the Council in several respects. First, the foreign ministers of Arctic states decided to establish a permanent Arctic Council secretariat to strengthen its capacity and provide institutional memory. Second, the ministers signed the first legally-binding agreement negotiated under the auspices of the Arctic Council. At the AC's 2009 meeting in Kautokeino, Norway, the United States delegation announced that the AC would serve as a forum to negotiate an agreement to "strengthen aeronautical and maritime search and rescue co-operation and coordination in the Arctic," with Russia co-sponsoring the proposal (Arctic Council, 2008, 2011: preamble and Article 7). Śmieszek and Koivurova (2017: 10) note that "because the Arctic Council has no independent legal personality" this agreement was between the eight Arctic states and not an "Arctic Council agreement" *per se*, with the Council instead serving "primarily as the catalyst for . . . negotiations and signature." This model was followed to reach the *Agreement on Cooperation on Marine Oil Pollution, Preparedness and Response in the Arctic* signed at the Ministerial in Kiruna, Sweden in May 2013. For many commentators, these binding agreements heralded the AC's evolution from a forum "set up to discuss, inform and potentially shape decisions by national governments" (Fenge and Funston, 2015: 10) towards a more formal decision-making function.

In 1999, the Arctic states debated whether to expand the AC's mandate to encompass economics issues, with Canada again insisting on a narrow mandate. The first two AC agreements contained an economic component, and a policymaker familiar with the "behind the scenes" machinations of the AC says that these treaties emerged from industry and insurance companies' requests for better safety planning (interview 5). For example, in 2012, Lloyds of London articulated that "the Arctic is likely to attract substantial investment over the coming decade, potentially reaching $100 [billion] or more," but that "the high risk/potentially high reward nature of Arctic investment" made firm predictions difficult (Emmerson *et al*., 2012). The Search and Rescue and Oil Pollution Prevention agreements affirm support for international law, create zones of responsibility for emergency response, provide resources to contact in case of emergency, allow states to ask for assistance in emergency response, and facilitate information exchange. Hence, Arctic states clarified areas of responsibilities and acted as facilitators to future human and commercial activities in the region. Such moves reduced uncertainty and ambiguity, placing heightened emphasis on pragmatic co-operation and clarity.

At the same time, these agreements dealt with relatively uncontroversial issues that reaffirmed areas of Arctic state jurisdiction and sovereignty. Public safety issues such as search-and-rescue operations and oil-spill preparedness had been the focus of state authorities since the early 2000s. Annual exercises held by national Arctic authorities, and more specifically coordinated by the national military organizations, focused on the responsiveness to such threats in their scenarios (Lackenbauer and Dean, 2016). Oil-spill preparedness arose as a particularly salient concern after the fiasco surrounding the 2010 British Petroleum oil spill in the Gulf of Mexico. Countries considered search-and-rescue operations vital because ice melt would increase access and bring more human activity to the Arctic region, whether by industry, tourists, or governments.

In 2017, the AC announced the *Agreement on Enhancing International Arctic Scientific Co-operation* (Arctic Council, 2017: article 4 and 5). David Balton, Deputy Assistant Secretary of the US Bureau of Oceans and International Environmental and Scientific Affairs, said: "There have been times in the past when permissions that scientists seek to do scientific research in another country have not been granted or not been granted as readily as need be" (State Department, 2017). The AC also yielded two "informal" agreements. The 2013 *Framework Plan for Cooperation on Prevention of Oil Pollution from Petroleum and Maritime Activities in the Marine Areas of the Arctic* offers a joint pledge to share information about oil pollution (Arctic Council, 2013b: 1.4.1), carry out joint pollution assessments (1.5.1), and develop coordinated best practices to deal with oil pollution (2.2.a). The 2013 *Framework for Action on Enhanced Black Carbon and Methane* saw states pledge to reduce black carbon and methane emissions (Arctic Council, 2013a). Clearly, the Council is a venue that states can use to create international agreements, either formal or informal in nature.

In sum, the Arctic states consider the AC the leading forum for regional governance and, accordingly, they have made the strongest commitment to this body. Since its creation, circumpolar governance through the AC has offered an important corrective to counterbalance narratives suggesting that the region is ungoverned or on the brink of conflict. The AC grew out of a desire to ease Cold War tensions and take collective action to address environment challenges. Where Nordic countries and to a lesser extent Canada sought a strong Council with a broad mandate, Russia and the United States argued successfully for a more limited forum which did not impinge upon state sovereignty. Although early attempts by the Nordic countries to use the AC as a venue to create policy failed, binding agreements negotiated under the auspices of the Council entrench a narrative of Arctic co-operation and collaboration supported by soft-law institutions rather than an overarching treaty. In addition to offering solutions to practical policy problems, the three binding agreements demonstrate the Council's role in facilitating these agreements between Arctic states and help to reinforce its place as the foremost regional governance body to a global audience.

The necessity of (limited) integration

Discussions about opening the regional dialogue to new stakeholders – non-state actors (such as private companies) or non-Arctic states – have raised fundamental questions about circumpolar governance and the primacy of the Arctic states.

Would denying new actors a place around the Arctic table undermine established governance regimes predicated on the primacy of Arctic states? An inclusive approach would provide opportunities for non-Arctic states and non-state actors to share their expertise, voice concerns, and exercise influence in the policy-making process. This inclusive philosophy embraces the "Global Arctic" concept, an analytical lens that conceptualizes how "the Arctic has become part and parcel of global political, economic, technological, environmental and societal change" (Heininen and Finger, 2018: 201). According to this logic, Arctic governance requires the inclusion of global players (particularly non-Arctic states) in key Arctic forums in order to "strengthen the decision-making process" (Grímsson quoted in Mazo, 2015: 247). The emergence of new platforms for a broader array of voices are either seen as a positive step in diversification that complements the AC, or a growing challenge to the Council's pre-eminence in regional governance.

The creation of the Arctic Circle Assembly (ACA) represented another such example. Its launch reflected a realignment emanating from tensions between Arctic state governments as well as growing pressures from non-Arctic state actors for a stronger voice in regional affairs. When Icelandic President Ólafur Ragnar Grímsson launched the ACA in 2013, the gathering was conceptualized as an open-tent venue in which diverse stakeholders with an interest in the Arctic region would partake in dialogues and exchanges on policymaking (Depledge and Dodds, 2016; Rossi, 2015). As such, the ACA allowed diverse types of stakeholders (NGOs, private sector, governments) to meet and debate policy issues, as well as to present findings by governments and academic researchers. With the big academic/policy conference format, Grímsson insisted that the ACA should be perceived as a supplement rather than a rival to the AC (Depledge and Dodds, 2016: 142–3). After all, the ACA lacked any pretence that it could create or enforce regional rules.

The "open-tent" messaging reflected a thinly veiled criticism of certain Arctic states (particularly Canada and Russia) that sought to limit the number of non-Arctic state observers in the AC, and who were perceived to desire the isolation of these stakeholders from critical venues of Arctic governance. Grímsson announced the launch of the ACA before the Kiruna Summit, when the Arctic states were slated to consider the observer applications of various non-Arctic states and non-state actors. Iceland and its foreign policy interests – particularly its relationship with China (with which it signed a free trade agreement in 2013) – aligned with accepting new observers (Depledge and Dodds, 2016: 142–4; Pan and Huntington, 2016). China developed significant bilateral and multilateral relations with Nordic countries in the late 2000s–early 2010s,[2] and Iceland signed commercial and scientific agreements with China in 2012. The two countries co-operated in launching the China-Iceland joint Aurora Observatory in 2016, and are engaged in joint ventures to explore the potential of oil and gas resources in the Dreki region (Lanteigne, 2017).

Analyzing the origins and types of participants of the ACA (see table 1.1) yields insight into the role that the assembly now plays in Arctic governance. Although the media and academic analysts tend to highlight the presence of non-Arctic states (especially China) at the annual event, careful scrutiny reveals an overwhelming

Table 1.1 ACA chairs, moderators, and speakers by home country during the 2013 and 2018 editions.

	2013	*2018*
Participants from Arctic states	73.2%	76.6%
Participants from non-Arctic European states	7.5%	11.8%
Participants from non-Arctic Asian states	12.6%	7.2%
Participants from other regions	0.8%	1.7%
Participants from global institutions (ex. UN)	5.9%	2.6%

Table 1.2 Provenances of ACA chairs, moderators, and speakers during the 2013 and 2018 editions.

	2013	*2018*
Academics, scientific researchers	41.4%	44.6%
National governments and bureaucrats	13.7%	17%
Companies	17.6%	11%
Civil society	16.8%	14.8%
Sub-national governments	3.5%	8.9%
International or regional organisations	3.1%	3.8%

dominance of speakers from the Arctic states in the ACA programming, reproducing their predominance in other forums such as the AC. Given that Arctic states fund Arctic research more heavily than other states, this is unsurprising. Asian states with advanced polar scientific research programs – Japan, Singapore, South Korea, and China – represent the bulk of Asian ACA contributors, whereas European contributors mainly originate from the European Union, the United Kingdom, Scotland, and Germany.

This data suggests that the global dimension of the ACA must be put in context, particularly the presence of participants from Asian states, who represent about 10% of interventions in ACA sessions. While the ACA offers opportunities for non-Arctic European and Asian participants to organize a few sessions, the Arctic states still dominate the proceedings. Breakout sessions specifically organized by non-Arctic states (Japan, the United Kingdom, and China), with speakers originating from the organizing country, are the exception. For example, three panels organized by Asian states accounted for 50% of all non-Arctic Asian speakers and chairs during the 2018 assembly.

Table 1.2 reveals that non-state actors clearly formed the majority of ACA speakers, chairs, and moderators. Academics and scientific researchers composed the most significant contingent of participants (see also Johannsdottir and Cook, 2017). As such, the ACA performs a generative function, facilitating the sharing of scientific findings from independent and government researchers to improve understandings of the Arctic environment and people. The presence of "marginalized" actors (Depledge and Dodds, 2016: 144), such as Alaska, Greenland, and the Faroe Islands, has increased over time. Indigenous leaders and non-governmental organizations, however, have remained underrepresented at the ACA – a significant shortcoming for a forum designed to diversify the Arctic dialogue.

For non-Arctic state officials, however, the ACA serves political purposes. Various national governments have sent delegations with the primary objective of framing or representing their respective states as having a "Northern" or "Arctic" nature. Presentations and sessions organized by the United Kingdom, Scotland, and China have stood out on this regard. As such, these sessions were at the crossroads of identity politics and political marketing, with each nation seeking to justify how they have contributed to the region and why they should play an enhanced role in Arctic governance. The national delegations reflect a deliberate effort to strike a balance between governmental officials and researchers, promoting scientific contributions as the main rationale for their involvement. For example, 75% of all speakers, chairs, and moderators from non-Arctic Asian countries who participated in the 2018 ACA were scientists or researchers, most of whom studied the natural sciences and presented their work on the regional and global environmental impacts of climate change. Other dominant themes related to economic and business development, sustainable development, and energy opportunities (Johannsdottir and Cook, 2017).

The annual ACA in Reykjavik has opened space for a range of Arctic stakeholders to create networks and influence regional agenda setting outside of the auspices of the Arctic Council and traditional bilateral and multilateral forums. The event has allowed participants to articulate diverse visions of the Arctic and suggest policies, norms, standards, and rules that could apply to regulate the region. Hence, the ACA is serving catalytic and generative functions with respect to Arctic governance. The popularity of the event since its inception, and the growing number of participants, reflects the significant value that Arctic stakeholders assign to it. Singapore, Québec, South Korea, the United States, and Greenland have organized smaller Arctic Circle Forums since 2015, indicating a perceived benefit in convening gatherings to target specialized issues and debates outside of the larger fall meeting in Iceland. While the ACA helped assemble diverse stakeholders, other forums developed to foster the participation of specific actors, such as the private sector.

The Arctic Economic Council

One of Canada's main efforts as chair of the AC from 2013–2015 centred around the creation of a circumpolar business forum – which ultimately became the Arctic Economic Council (AEC) – to build circumpolar trade and foster business and natural resource development opportunities in the Far North. Its overarching theme, "Development for the People of the North," and its three sub-themes (Responsible Arctic Resource Development, Safe Arctic Shipping, and Sustainable Circumpolar Communities) all connected with the economic development focus of its domestic Northern Strategy, and reflected why private-sector actors should have more input into AC processes (Exner-Pirot, 2016; Lackenbauer, 2017). However, Canada, "encountering significant resistance from representatives favoring a more environmentally conscious approach to such economic development efforts," insisted that "the Council needed to get on board with the proposed plan and not engage in unnecessary debate and delay" (Nord, 2017: 40).

While debates continued in Canadian academic and advocacy circles over the role of non-renewable resource development as a backbone of a sustainable Northern economy (e.g. Everett and Nicol, 2014; Harrington and Lecavalier, 2014; Nicol, 2015), the Conservative government under Stephen Harper had shown a clear commitment to prioritizing this form of development as the conduit to improve socio-economic conditions in Canada's Arctic. Minister Leona Aglukkaq, who served as the AC chair, insisted that the forum needed "better engagement of the business community, . . . to talk to the people that live in the North, that are developing the North, that are shipping through the North, that are mining in the North" (Aglukkaq cited in Koring, 11 May 2013).

AC member states, who created the Circumpolar Business Forum Task Force during the Swedish AC chair (2011–2013), backed the Canadian initiative. The AEC held its first meeting in Iqaluit in September 2014. As an "independent organization that facilitates Arctic business-to-business activities and responsible economic development through the sharing of best practices, technological solutions, standards and other information," the AEC's membership represents a wide range of business interests: "from mining and shipping companies to reindeer herding and Aboriginal Economic Development Corporations" (AEC, 2016a). The new business forum also mirrored some key structures of the AC, with member states and permanent participants enjoying privileged status in the new body. All Arctic member states could name up to three representatives to represent their interests at the AEC, for a total of 42 possible representatives.

The place assigned to the AEC in Arctic governance lacked sufficient detail, raising various concerns about its potential to undermine the AC and the unique status of permanent participants, or to provide "transnational corporations with preferential access to national governments" (Axworthy and Simon, 2015; also Exner-Pirot, 2016: 89–90, 93). It was unclear if the new body would have a close working relationship with the AC, and, if so, how this co-operation would materialize (Young, 2016: 110). In practice, the private sector actors involved in the AEC have demonstrated a strong awareness of and interest in finding their niche in a broader global governance landscape, including the AC and beyond. For example, mentions of global institutions are present in the different work plans and mandates of the AEC's working groups, such as references to the United Nations Sustainable Development Goals and the World Economic Forum (WEF)'s Arctic Investment Protocol (AIP) as key initiatives in the Energy and Investments and Infrastructure working groups.

The AEC's formal relationship with the Council itself, however, remained largely undeveloped after Canada turned over the AC chairmanship to the United States in 2015, and no AEC co-operation with the AC or its working groups ensued for the following two years (Chater, 2017). Although Canadian delegates attempted to draw attention to economic issues and sought to clarify the AC's relationship to the AEC during the American chair, Senior Arctic Officials (SAOs) made it clear that they were not interested in further discussions about economic issues at that time (Arctic Council, 2016: 25). A focus on business issues was revived under the Finnish Chairmanship (2017–2019), when AEC chair Tero Vauraste became the first business actor to provide an update/briefing to the AC. He emphasized the importance of global market links and public-private partnerships for pursuing Arctic sustainable economic development, while expressing "hope to see a growing contingent of people from the business community engaged with the Arctic Council's proceedings" (Arctic Council, 2018: 21; see Chapter 5 for an in-depth discussion of AEC/AC relations).

The AEC's strategic plan for 2019–2021 coincides with the Icelandic chair of the AC and aligns with Iceland's focusing on economic development issues and building ties with a diversified range of stakeholders – particularly non-Arctic states. While the previous AEC strategic plan's first overarching theme sought to build "strong market connections between the Arctic states" (AEC, 2016b), the new plan highlights the need to foster stronger connections with global value chains (AEC, 2019). Moving beyond an inward focus, this suggests a deepening awareness of the limitations to fixating on regional economic dynamics at the expense of global forces that influence, shape, and in some cases dictate the form and pace of Arctic development.

Conclusion

The regional governance initiatives put forward by the Arctic states have changed in three major ways. First, the Arctic Council (AC) began as a limited environmental governance body created and dominated by the Arctic states. Disagreements about the Council's role in policy-making and the elasticity of its mandate produced a stronger AC. Second, the policy-making role and mandate of the AC has expanded, with various international agreements negotiated under its auspices in the past decade. Third, new circumpolar forums such as the Arctic Circle Assembly (ACA) and Arctic Economic Council (AEC) have diversified discussions about Arctic governance, allowing a broader array of actors to voice their opinions.

Regional governance infrastructure remains fragmented and revolves around the preeminent status that Arctic states have carved up for themselves, with the Arctic coastal states' predominant role in aspects of Arctic Ocean governance directed by the international Law of the Sea (see Chapter 2 for more on this). The AC was created in 1996 to serve as a co-operative forum on environmental protection and social development, and the *ad hoc* nature of this initiative shaped circumpolar governance for more than a decade. Despite its relatively weak structure and funding, the AC offered a collaborative platform for Arctic states, in close partnership with Indigenous Permanent Participants, to launch meaningful initiatives. The 2004 *Arctic Climate Impact Assessment* (discussed in Chapter 4) is a striking example of how converging interests among Arctic states led to a collective endeavour that reshaped both regional and global understandings about climate change. The same logic applies to the search-and-rescue and oil-spill preparedness agreements struck at the AC after 2011. Both agreements reiterated the dominant role of Arctic states in regional governance, essentially reinforcing and harmonizing national legislation and jurisdiction in a manner consistent with state sovereignty.

Although Arctic states still dominate Arctic governance, limited integration of non-Arctic stakeholders has emerged in recent years. This opening of the "Arctic tent" has led to a diversification of stakeholders involved in circumpolar discussions – albeit with non-Arctic states and non-state actors assigned a secondary or marginal role. While forums like the ACA have offered a more public role for non-Arctic states, their participation remains modest. In the future, however, they are likely to heighten the pressure for an increasing role in Arctic governance. The traditional emphasis on national jurisdictions and sovereignty made sense when anticipated activities were confined to the territories and exclusive economic zones of Arctic states. With the acceleration of global warming, the prospective opening of the Central Arctic Ocean (an international space) to human activities introduces new practical concerns about shipping, resources, and environmental protection that will require a global approach. The moratorium on fisheries in the Central Arctic Ocean, explored in the next chapter, provides an indication of how regional governance is changing to accommodate the interests of Arctic states and non-Arctic states as rightsholders.

Notes

1 Although the public face of the AC conveyed the image of a collegial body that did not inspire rivalries among its members, not all deliberations went smoothly. United States officials believed that the Russian government re-directed some funds from the United Nations' Global Environmental Facility, intended for environmental protection, to non-environmental projects (interview 10).

2 The China-Nordic Arctic Research Council would be a noteworthy example.

References

Arctic Council. 1998. *The Iqaluit Declaration*. Tromsø, Norway: Arctic Council Secretariat, 1998.

Arctic Council. 1999. *Senior Arctic Officials Meeting Highlights, November 17–19, 1999, Washington, D.C., U.S.A.* Tromsø, Norway: Arctic Council Secretariat.

Arctic Council. 2008. *Meeting of Senior Arctic Officials Final Report, 19–20 November 2008, Kautokeino, Norway*. Tromsø, Norway: Arctic Council Secretariat.

Arctic Council. 2011. *Agreement on Co-operation on Aeronautical and Maritime Search and Rescue in the Arctic*. Nuuk, Greenland: Arctic Council.

Arctic Council. 2013a. *Enhanced Black Carbon and Methane Emissions Reductions: An Arctic Council Framework for Action*. Tromsø, Norway: Arctic Council Secretariat, 2013.

Arctic Council. 2013b. *Framework Plan for Co-operation on Prevention of Oil Pollution from Petroleum and Maritime Activities in the Marine Areas of the Arctic*. Tromsø, Norway: Arctic Council Secretariat.

Arctic Council. 2016. *Summary Report, SAO Plenary Meeting. Fairbanks, Alaska, 16–17 March 2016*. Tromsø, Norway: Arctic Council Secretariat.

Arctic Council, 2017. *Agreement on Enhancing International Arctic Scientific Co-operation.* Tromsø, Norway: Arctic Council Secretariat.

Arctic Council. 2018. *Report: SAO Plenary Meeting, Rovaniemi, Finland, 1–2 November 2018.* Tromsø, Norway: Arctic Council Secretariat.

Arctic Economic Council. 2016a. "Backgrounder." http://arcticeconomiccouncil.com/about-us/backgrounder/.

Arctic Economic Council. 2016b. *Strategic Planning Document, 2016–2018.* April 27, 2016. Tromsø, Norway: Arctic Economic Council Secretariat.

Arctic Economic Council. 2019. *Strategic Planning Document 2019–2021.* Tromsø, Norway: Arctic Economic Council Secretariat.

Axworthy, Lloyd and Mary Simon. 2015. "Is Canada Undermining the Arctic Council?" *Globe and Mail*, March 4.

Bloom, Evan T. 1999. "Establishment of the Arctic Council." *The American Journal of International Law* 93, no. 3, 712–722.

Blunden, Margaret. 2009. "The New Problem of Arctic Stability." *Survival* 51, no. 5: 121–142.

Byers, Michael. 2012. *Circumpolar Challenges: An Ambitious Agenda for the Arctic Council.* Ottawa, Ontario: Rideau Institute.

Chater, Andrew. 2017. "Understanding Canada's Role in the Evolution of the Arctic Council." In *One Arctic: The Arctic Council and Circumpolar Governance*, edited by P. Whitney Lackenbauer, Heather Nicol, and Wilfrid Greaves, 78–100. Ottawa, Ontario: Canadian Arctic Resources Committee.

Depledge, Duncan and Klaus Dodds. 2016. "Bazaar Governance: Situating the Arctic Circle." In *Governing Arctic Change*, edited by Kathrin Keil and Sebastian Knecht, 141–160. London: Palgrave Macmillan.

Dodds, Klaus. 2011. "We Are a Northern Country: Stephen Harper and the Canadian Arctic." *Polar Record* 47, no. 4: 371–374.

Emmerson, Charles *et al.* 2012. *Arctic Opening: Opportunity and Risk in the High North.* London: Chatham House and Lloyds.

English, John. 2013. *Ice and Water: Politics, Peoples and the Arctic Council.* Toronto, Ontario: Allen Lane.

Everett, Karin and Heather Nicol. 2014. "Economic Development, Indigenous Governance, & Arctic Sovereignty." *Arctic Yearbook* 3: 1–19.

Exner-Pirot, Heather. 2016. "Canada's Arctic Council Chairmanship (2013–2015): A Post-Mortem." *Canadian Foreign Policy Journal* 22, no. 1: 84–96.

Fenge, Terry and Bernie Funston. 2015. *The Practice and Promise of the Arctic Council.* Toronto, Ontario: Greenpeace Canada

Gordon Foundation. 2010. *A Brief History of the Creation of the Arctic Council.* Toronto, Ontario: Gordon Foundation.

Harrington, Cameron and Emma Lecavalier. 2014. "The Environment and Emancipation in Critical Security Studies: The Case of the Canadian Arctic." *Critical Studies on Security* 2, no. 1: 105–119.

Heininen, Lassi and Matthias Finger. 2018. "The 'Global Arctic' as a New Geopolitical Context and Method. *Journal of Borderlands Studies* 33, no. 2: 199–202.

Huebert, Rob. 1998. "New Directions in Circumpolar Co-operation: Canada, the Arctic Environmental Protection Strategy and the Arctic Council." *Canadian Foreign Policy* 5, no. 2, 37–57.

Interview 1 with Ray Arnaudo, former United States delegation member and Council chair, April 17, 2013.

Interview 5 with former Canadian delegation member and Yukon government employee, winter 2013.
Interview 7 with former United States Council delegation member and current environmental official, spring 2013.
Interview 8 with former United States Council delegation member and current environmental research official, spring 2013.
Interview 9 with former United States delegation member and government of Alaska official, spring 2013.
Interview 10 with former United States delegation member and EPA employee, spring 2013.
Interview 14 with former United States high-ranking delegation member and State Department official, spring 2013.
Interview 15 with Chester Reimer, Inuit Circumpolar Council consultant and delegation member, February 13, 2013.
Interview 16 with Sauli Rouhinen, former Finnish SDWG chair, May 30, 2013.
Interview 18 with Russian government environmental researcher and former delegation member, summer 2013.
Johannsdottir, Lara and David Cook. 2017. "Discourse Analysis of the 2013–2016 Arctic Circle Assembly Programmes." *The Polar Record* 53, no. 270: 276–279.
Keskitalo, Eva Carina Helena. 2004. *Negotiating the Arctic: The Construction of an International Region*. New York: Routledge.
Koivurova, Timo. 2008. "Alternatives for an Arctic Treaty: Evaluation and a New Proposal." *Review of European, Comparative and International Environmental Law* 17, no. 1: 14–26.
Koivurova, Timo. 2010. "Limits and Possibilities of the Arctic Council in a Rapidly Changing Scene of Arctic Governance." *Polar Record* 46, no. 2: 146–156.
Koivurova, Timo and David L. Vanderzwaag. 2006. "The Arctic Council at 10 Years: Retrospect and Prospects." *UBC Law Review* 40, no. 1: 121–194.
Koring, Paul. 2013. Leona Aglukkaq's Personal Arctic Stake. *Globe and Mail*, May 11.
Lackenbauer, P. Whitney. 2009. "From Polar Race to Polar Saga: An Integrated Strategy for Canada and the Circumpolar World." *Foreign Policy for Canada's Tomorrow No. 3*. Toronto, Ontario: Canadian International Council.
Lackenbauer, P. Whitney. 2017. "Conceptualizing 'One Arctic' as the 'Canadian Arctic'? Situating Canada's Arctic Council Chairmanship (2013–15)." In *One Arctic: The Arctic Council and Circumpolar Governance*, edited by P. Whitney Lackenbauer, Heather Nicol, and Wilfrid Greaves, 46–78. Ottawa: Canadian Arctic Resources Committee.
Lackenbauer, P. Whitney and Ryan Dean. 2016. "Canada's Northern Strategy Under the Harper Conservatives: Key Speeches and Documents on Sovereignty, Security, and Governance, 2006–15." *Documents on Canadian Arctic Sovereignty and Security (DCASS) No. 6*. Calgary, Alberta and Waterloo, Ontario: Centre for Military, Strategic and Security Studies/Centre on Foreign Policy and Federalism/Arctic Institute of North America.
Landriault, Mathieu. 2016. "Interest in and Public Perceptions of Canadian Arctic Sovereignty: Evidence from Editorials, 2000–2014." *International Journal of Canadian Studies* 54, no. 1: 5–25.
Lanteigne, Marc. 2017. "'Have You Entered the Storehouses of the Snow?' China as a Norm Entrepreneur in the Arctic." *Polar Record* 53, no. 269: 117–130.
Mazo, Jeffrey. 2015. "Showing the Flag." *Survival* 57, no. 1: 241–252.

Munk-Gordon Arctic Security Program. 2012. *Canada as an Arctic Power: Preparing for the Canadian Chairmanship of the Arctic Council 2013–2015*. Toronto, Ontario: Munk School of Global Affairs and the Walter and Duncan Gordon Foundation.

Nicol, Heather. 2015. "Canada's Arctic Agenda: Failing to Make a Case for Economic Development as an International Strategy in the Circumpolar North?" In *The Handbook of the Politics of the Arctic*, edited by Leif Christian Jensen and Geir Hønnelan, 51–65. Cheltenham, UK: Edward Elgar.

Nord, Douglas. 2017. "Leadership from the Chair: The Experience of Three Successive Chairmanships of the Arctic Council." In *One Arctic: The Arctic Council and Circumpolar Governance*, edited by P. Whitney Lackenbauer, Heather Nicol, and Wilfrid Greaves, 27–45. Ottawa: Canadian Arctic Resources Committee.

Pan, Min and Henry Huntington. 2016. "A Precautionary Approach to Fisheries in the Central Arctic Ocean: Policy, Science, and China." *Marine Policy* 63: 153–157.

Rossi, Christopher. 2015. "The Club Within the Club: The Challenge of a Soft Law Framework in a Global Arctic Context." *The Polar Journal* 5, no. 1: 8–34.

State Department. 2017. "Briefing on the Arctic Council Ministerial." May 8. www.state.gov/r/pa/prs/ps/2017/05/270728.htm.

Śmieszek, Małgorzata (Gosia) and Timo Koivurova. 2017. "The Arctic Council: Between Continuity and Change." In *One Arctic: The Arctic Council and Circumpolar Governance*, edited by P. Whitney Lackenbauer, Heather Nicol, and Wilfrid Greaves, 1–26. Ottawa: Canadian Arctic Resources Committee.

United States. 1983. "United States Arctic Policy." *National Security Decision Directive 90*. April 14. http://fas.org/irp/offdocs/nsdd/nsdd-090.htm.

Wegge, Njord. 2012. "The EU and the Arctic: European Foreign Policy in the Making." *Arctic Review* 3, no. 1: 6–29.

Young, Oran R. 1998. *Creating Regimes: Arctic Accords and International Governance*. Ithaca, NY: Cornell University Press.

Young, Oran R. 2005. "Governing the Arctic: From Cold War Theater to Mosaic of Cooperation." *Global Governance* 11, no. 1: 9–15.

Young, Oran R. 2016. "The Arctic Council at Twenty: How to Remain Effective in a Rapidly Changing Environment." *UC Irvine Law Review* 6, no. 2: 99–120.

2 Regional governance of Arctic Ocean

The Arctic 5 (A5) and the Arctic 5+5

As the season for open water extends concomitant to declining sea ice in the Arctic Ocean, so too do economic opportunities associated with shipping, tourism, offshore oil and gas exploration, and commercial fishing. These activities (both real and anticipated) raise concerns, particularly in the scientific and environmental communities, about a perceived rush to exploit natural resources that may threaten marine ecosystems and affect traditional ways of life of Indigenous peoples and northern communities.

Circumpolar politics have tended to fixate on national jurisdictions and issues within the territorial seas or exclusive economic zones of the Arctic Ocean coastal states – commonly designated as the "Arctic 5" or "A5." International law confers on coastal states unique rights, obligations, and interests which, the A5 believe, give them primary responsibility for managing the Arctic Ocean. Members of the A5 insist that, in setting up the Arctic Council (AC), no one intended to preclude Arctic states from meeting in other settings to discuss areas of mutual interest or responsibility. As this chapter discusses, other AC members (Iceland, Finland, and Sweden) and permanent participants have been highly critical when they have been left out of A5 discussions, as have a growing number of non-Arctic state and non-state actors. While other groupings of Arctic countries meet regularly (such as the Barents Council and the Nordic Council) without causing concerns, the A5 meeting format raises questions about whether the coastal states seek to assert their "unique position" to shut out other stakeholders and rightsholders in Arctic Ocean governance.

This chapter charts the emergence of A5 meetings in response to pessimistic narratives suggesting that the Arctic lacked the legal basis for stable governance and anticipated conflict between Arctic states over boundaries, continental shelves, and resources. Through the *Ilulissat Declaration* in 2008, the A5 sought to reassure the international community that existing governance regimes and an extensive legal framework provided a firm basis for responsible regional management by the Arctic coastal states – and reasserted their sovereign rights over parts of the Arctic Ocean and the continental shelves that lay beneath. This messaging emanated from their legal position as coastal states adhering to the United Nations Convention on the Law of the Sea (UNCLOS) – although, in the case of the United States, not as a party to the convention but as customary international law.

While Arctic coastal regions rely heavily on domestic legislation and policy frameworks, about one-fifth of the Arctic Ocean (2.8 million square kilometers) lies beyond national jurisdiction and lacks comprehensive and efficient instruments for the conservation and sustainable use of marine biodiversity (De Lucia *et al.*, 2018). Accordingly, the melting and prospective opening of the Arctic Ocean introduces a new chapter into regional governance in which areas of high seas will be a focal point. The Central Arctic Ocean (CAO) area that lies beyond national jurisdiction calls upon governance mechanisms beyond purely regional ones, in which Arctic and non-Arctic stakeholders must be involved to achieve enforceable norms, rules, and standards for this area. Working through the specific issue of future Arctic fisheries beyond the Exclusive Economic Zones (EEZs) of the Arctic coastal states, the second half of this chapter shows how and why an "Arctic 5+5" format emerged to generate regional norms and rules.

The Arctic 5 (A5): meetings of the Arctic Ocean coastal states

In the mid-2000s, a growing throng of alarmist commentators challenged the post-Cold War narrative of circumpolar peace and co-operation by suggesting that climate change would be the catalyst for Arctic conflict, tied to an alleged "race for resources" as the world scrambled to claim the riches of a newly accessible region. Over-hyped controversies such as the Hans Island dispute between Canada and the Kingdom of Denmark; longstanding disagreements between the United States and both Canada and Russia over the legal status of waters in the Northwest Passage and Northern Sea Route respectively; the undefined Norwegian-Russian boundary in the Barents Sea; and the prospect of overlapping extended continental shelf claims in the Arctic Ocean were mobilized to highlight regional uncertainty and friction. Environmental and political developments in mid-2007 affirmed this framework of impending polar peril. First, the *Arktika* scientific expedition placed a titanium Russian flag on the seabed at the geographic North Pole, leading Canada's Foreign Affairs Minister Peter MacKay to accuse the Russians of laying claim to the Arctic in the style of fifteenth-century imperialists. Although the Russian government took immediate action to try to dispel Western concerns that this action represented anything resembling a territorial claim, mistrust lingered (Livermore, 2017: 54–6). Record lows in the extent of Arctic sea ice later that year affirmed what the *Arctic Climate Impact Assessment* report portended a few years before: that the cryosphere was melting fast and reshaping the polar world. In early 2008, Scott Borgerson's *Foreign Affairs* article anticipating an "Arctic Meltdown" became an international sensation, envisaging the prospect of armed conflict over energy and mineral resources in the region. On the pretext that the Arctic was an emerging frontier without an overarching government, various commentators began to assert an urgent need for new forms of governance, including appeals for a comprehensive, legally binding Arctic treaty (e.g. Huebert and Yaeger, 2008; Huebert, 2009; Hertell, 2009; Koivurova and Molenaar, 2009).

To counter the popular misperception that the Arctic was a "lawless frontier" and a "reincarnation of the Wild West" (Riddell-Dixon, 2008), high-level political representatives from the five Arctic Ocean coastal states (Canada, the Kingdom of Denmark, Norway, the Russian Federation, and the United States) convened their first formal meeting in Ilulissat, Greenland, in May 2008 to discuss legal regiments and jurisdictional issues. The ensuing *Ilulissat Declaration* highlighted that an extensive international legal framework already applied to the Arctic Ocean (notably the law of the sea), and that the A5 shared a unified commitment to the orderly settlement of any possible overlapping claims to the extended continental shelf. Although the discussions touched on the Arctic Ocean, climate change, the protection of the marine environment, and maritime safety – all issues of core interest to the Arctic Council – the decision to exclude the three Arctic states without Arctic Ocean coastlines (Iceland, Finland, and Sweden) as well as the Indigenous permanent participants reflected a deliberate attempt to focus on coastal state rights to counter calls for a new comprehensive international legal regime to govern the Arctic Ocean. The *Ilulissat Declaration* (2008) asserted that:

> By virtue of their sovereignty, sovereign rights and jurisdiction in large areas of the Arctic Ocean the five coastal states are in a unique position to address these possibilities and challenges. . . . This framework provides a solid foundation for responsible management by the five coastal States and other users of this Ocean through national implementation and application of relevant provisions. We therefore see no need to develop a new comprehensive international legal regime to govern the Arctic Ocean.

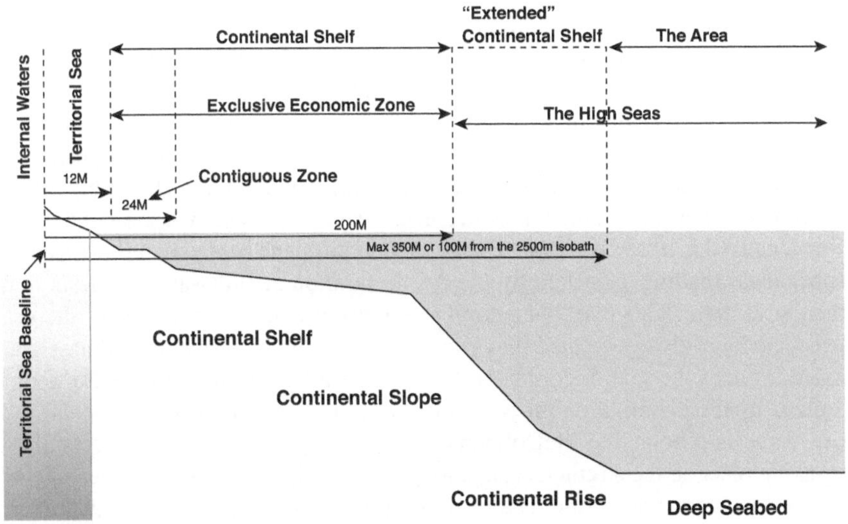

Figure 2.1 Maritime Boundaries under UNCLOS

Source: Global Affairs Canada

Through the declaration, the A5 sent a strong political signal that they supported the existing global framework provided by "the law of the sea [as] a solid foundation for responsible management by the five coastal States." This stance is predicated on the idea that an extensive international legal framework applies to the Arctic Ocean as it does to other maritime regions around the world. Under the 1982 United Nations Convention on the Law of the Sea (UNCLOS), or "constitution of the oceans," Arctic coastal states have specific rights and obligations that cast them in a central role in regional ocean management. For example, under article 76(1) of the Convention, coastal states are entitled to claim a continental shelf up to 200 nautical miles from their territorial sea baseline, which is also the maximum extent of the Exclusive Economic Zone (EEZ). Under the legal regime governing the EEZ, coastal states have sovereign rights over the natural resources of the water column and the seabed, as well as jurisdictional authority over certain specific matters like marine scientific research and the protection and preservation of the marine environment. Further, coastal states have sovereign rights over the natural resources of the seabed and subsoil of their continental shelf both within and beyond 200 nautical miles, as well as jurisdictional authority over certain activities (such as marine scientific research).[1]

UNCLOS sets out a process for coastal states to determine the precise limits of their extended continental shelf. This process involves making a submission to an expert body called the Commission on the Limits of the Continental Shelf (CLCS) which reviews the submission, assesses the extent to which a state has defined its outer continental shelf in conformity with the technical scientific requirements laid out in article 76 of the Convention, and makes recommendations to the state. (Jares, 2009; Reichert, 2009; McDorman, 2010). UNCLOS provides that limits established by the coastal state on the basis of the CLCS's recommendations are final and binding (Elferink and Johnson, 2006), and the Commission does not apportion shelf between states. Areas in dispute or claimed by more than one state are outside its remit. As a result, any overlaps of continental shelf between coastal states must be resolved bilaterally through negotiation or third-party adjudication (such as arbitration or submitting the matter to the International Court of Justice or the International Tribunal for the Law of the Sea) (Global Affairs Canada, 2016).

Although popular commentaries suggested that the different Arctic coastal states were engaged in a "scramble" or a "race" to gobble up the continental shelf, a more sober legal reading – backed by the A5 declaration at Ilulissat – suggests that their sovereign rights over the natural resources of extended continental shelves exist and have always existed by virtue of state sovereignty over a landmass that extends under the sea. Accordingly, those rights are in no way affected or jeopardized by the delimitation process, nor changed by the melting of the polar ice cap. As legal scholar Ted McDorman (2010) astutely observed, "what 'scramble' is taking place in the Arctic Ocean amongst the bordering States has been one of seeking to acquire scientific data respecting the geologic composition and other physical properties of the continental margin areas in the Arctic Ocean" (195–6). Accordingly, Arctic coastal states have invested heavily in scientific, technical, and legal activities and assessments necessary to prepare their submission to the

CLCS, often in close collaboration with one another. Even though the outer limits of the continental shelves of these countries are expected to overlap, collaborative seabed-mapping brings mutual scientific benefit and may even serve as a "foundation for joint ecosystem-based, integrated management" of ocean areas (Baker, 2009; also Baker and Yeager, 2015).

Putting together these applications is careful, time- and resource-consuming work. The applications are then handled in a process that is lengthy, in part due to capacity challenges within the receiving body. The CLCS has encountered financial difficulties in terms of fulfilling all the experts' positions and ensuring their presence at the Commission to proceed with the technical analyses. As of December 2018, 81 separate submissions have been filed at the CLCS; recommendations had been adopted for only 29 submissions, while technical sub-committees have been struck for an additional 12 submissions. This means that no work has been done for 40 submissions (CLCS, 2018); limited funding has greatly affected the efficiency of the CLCS. Member states at the United Nations have been unwilling to equip the body with sufficient means to carry its mandate in an efficient manner. Institutional shortcomings are likely to affect the tempo of this so-called race.

The idea that outstanding Arctic boundary disputes would be solved through diplomacy received a boost in April 2010, when Russia and Norway resolved a 40-year disagreement over the division of the Barents Sea. Cajoling other Arctic coastal states to note this landmark resolution, Sergei Lavrov and Jonas Gahr Støre (the Russian and Norwegian foreign ministers respectively) explained that

> the Law of the Sea provided a framework that allowed us to overcome the zero-sum logic of competition and replace it with a process focused on finding a win-win solution. We hope that the agreement will inspire other countries in their attempts to resolve their maritime disputes, in the High North and elsewhere, in a way that avoids conflict and strengthens international co-operation.
>
> (Lavrov and Støre, 2010)

Given the vested interests of the Arctic coastal states to assert their sovereign rights under international law and to sustain a narrative of regional peace and stability, Canadian foreign minister Lawrence Cannon hosted a second meeting of A5 foreign ministers in Chelsea, Quebec in late March 2010. The organizers emphasized that these meetings complemented rather than undermined the Arctic Council (AC)'s work by identifying areas of common interest that fell within the countries' respective sovereign jurisdictions, laying the groundwork for states to translate ideas generated by the AC and other soft-law bodies into concrete policy. Accordingly, the chair's statement highlighted the importance of coastal state collaboration on continental shelf delineation, Arctic search and rescue, a mandatory shipping regime in polar waters, the creation of an Arctic Regional Hydrographic Commission, and further scientific research into Arctic fisheries. "We had a forward-looking dialogue on issues related to our roles and responsibilities in areas under our jurisdiction in the Arctic Ocean," Cannon noted. "This was in furtherance of commitments in the 2008 *Ilulissat Declaration* and complements our discussions and collaboration with all Arctic states, Arctic indigenous peoples and others through the Arctic Council, the central forum for international co-operation on Arctic issues, and other fora" (Cannon, 2010).

Despite reassuring messaging that sought to situate the A5 format within the broader circumpolar governance mosaic, the Chelsea meeting generated significant criticism from various sectors. Iceland, Finland, and Sweden argued that A5 meetings threatened to undermine the role of the AC as the premier forum for Arctic co-operation. The European Union (EU) opposed what it saw as the narrowing of the Arctic agenda, and the United States seemed reticent about the A5 format. "Significant international discussions on Arctic issues should include those who have legitimate interests in the region," Secretary of State Hillary Clinton told reporters – a statement that law professor Betsy Baker saw less as a repudiation than a "friendly reminder of the need to support the Arctic Council" (quoted in CBC News, 2010). Permanent participants were critical of being excluded from these meetings as Indigenous rightsholders with a primary responsibility to ensure responsible and sustainable Arctic development and stewardship. For example, Duane Smith, the president of the Inuit Circumpolar Conference (Canada), argued that Inuit needed direct representation as "a coastal people, because this summit is about the Arctic Ocean coast, and because Mr. Cannon underlined the importance of our involvement in multilateral meetings outside the Arctic Council" (quoted in Boswell, 2010).

Since the 2010 Chelsea ministerial, senior officials from the A5 have hosted various meetings to further discuss fisheries, energy, and public safety issues. Furthermore, scientists and officials from the Arctic coastal states have met to discuss issues related to the continental shelf. The political profile of these meetings, however, has declined. Even if the A5 might want to manage the Arctic Ocean as a private sea, the Arctic coastal states recognize that they lack the sovereign rights or jurisdiction in the water column beyond 200 nautical miles or over the entire Arctic seabed.[2] Arctic states have dismissed the idea of an Arctic Treaty along the lines of the Antarctic one, emphasizing core differences between the southern continent and an Arctic Ocean surrounded by sovereign states with clear rights conferred under established international law, instead preferring what political scientist Oran Young described as a "messier . . . governance complex" that seeks to "devise issue-specific regulative regimes to address concerns involving shipping, fishing, and off-shore oil and gas development" (Young, 2009).

Fisheries: from domestic action to the A5 to the A5+5

Fisheries in the Arctic Ocean are both uncertain and limited, with about 20% of the Central Arctic Ocean (CAO) area lying beyond national jurisdiction. Current general knowledge of Arctic marine ecosystems remains limited, with the multi-year ice pack precluding systematic research into fish species distribution and abundance, northern colonization by fish species, and the potential ecosystem effects of fishing. Furthermore, scientific knowledge is scant about the effects of climate change and ocean acidification on the oceanography and primary productivity of the Arctic Ocean. With the warming of sub-Arctic oceans, however, it is likely that certain fish stocks will migrate north in search of cooler waters (Molenaar, 2012; Hollowed *et al.*, 2013).

Based on the available information, most of the scientific community anticipates that changes necessary to support sustainable populations of commercially viable fish stocks on the high seas of the Arctic Ocean (particularly the CAO) will occur slowly, if at all.[3] Nevertheless, environmental groups have expressed concern that major non-Arctic fishing nations (e.g. most member States of the EU, China, South Korea, and Japan) may seek to exploit fishery resources in the CAO (e.g. Solli *et al.*, 2013). Because these waters are high seas, and thus beyond the national jurisdiction of the Arctic coastal states, they would be within their rights to do so under global instruments (such as UNCLOS, the 1995 UN Fish Stocks Agreement [UNFSA], and the 1995 Code of Conduct for Responsible Fisheries).

For its part, the Arctic Council has had no direct management role in fisheries, and has no mandate or capacity to act as a regional fisheries organization. Accordingly, it should not be misrepresented as an existing or potential regional fisheries management organization or arrangement (RFMO/A).[4] As an intergovernmental forum, however, it can serve as a useful avenue to promote research and facilitate the exchange of scientific information on fish stocks and marine ecosystems. Typical international practice is to establish an RFMO/A based on historical and existing fishing activity, but there is no fishing history in the CAO due to persistent ice cover. Accordingly, the high seas portion of the CAO is not subject to a comprehensive regional fisheries agreement, and only a southern portion of the CAO (north of Greenland and Svalbard) fall within the area covered by the North-East Atlantic Fisheries Commission. Other relevant RFMO/As lack a geographical or substantive mandate for comprehensive fisheries regulation in the CAO.

By extension, fundamental questions emerge about who should be engaged in fishery governance, from the national scale, to the circumpolar regional level, to global rights in the CAO. Pursuant to UNCLOS and the UNFSA, coastal states have responsibilities to manage living resources in their respective EEZs, including straddling fish stocks, and thus have authority to control fishing activities within these waters. For example, the United States implemented a moratorium on new commercial fisheries in its Arctic EEZ in December 2009 (NOAA, 2009). This supported a Senate resolution directing the government to negotiate a new international agreement for managing migratory and transboundary fish stocks in the Arctic Ocean. The moratorium covers finfish, molluscs, crustaceans, and other forms of marine animal and plant life other than marine mammals and birds. (It does not apply to Pacific salmon and Pacific halibut, which are managed under other legal authorities, or to sport fishing and subsistence fishing/hunting within territorial waters.) This initiative was a precautionary measure, as there was no commercial fishing in the moratorium area in 2009. The US desire to secure a pre-emptive moratorium was primarily rooted in concerns about potential unregulated fishing of migratory, transboundary, and straddling stocks on the high seas bordering the Bering Sea and to avoid its experience in the North Pacific high seas, where the pollock fishery collapsed before a 1994 regulatory regime was put in place (Bailey, 2011; Zou and Huntington, 2018).

The United States claimed the right, as a coastal state, to declare a moratorium within its territorial sea and EEZ (although in this case it did have international implications because of the disputed maritime boundary with Canada in the Beaufort Sea). Other countries have taken similar unilateral action. In October 2014, for example, the Canadian federal government announced that it would not permit new commercial fishing in the Beaufort in advance of further scientific research. Bilaterally, a shared interest in preserving living marine resources prompted Canadian Prime Minister Justin Trudeau to declare in a March 2016 joint statement with President Obama that

> Canada commits to working with Northern and Indigenous communities to build world-leading and abundant Arctic fisheries – based on science – that firstly benefit Northern communities. Together, the United States' and Canada's actions will create the largest contiguous area of well-regulated fisheries in the world.
>
> (Government of Canada, 2016)

The next step elevated the fisheries issue to the Arctic coastal state level, given the countries' clear legal rights to manage fish stocks in their internal waters, territorial waters, and EEZs. In June 2010, the A5 met in Oslo, Norway to discuss coastal state responsibilities concerning conservation and management of fish stocks in the Arctic Ocean, where they agreed upon the need for further scientific research on the state and nature of fish stocks and Arctic Ocean ecosystems. Consequently, the First Meeting of Scientific Experts on Fish Stocks in the Central Arctic Ocean (FiSCAO) was held in Anchorage, Alaska (United States) in June 2011. Participants observed that "even though there is much useful information available regarding climate change and other changes in the Arctic marine ecosystems," they identified "substantive gaps in information and understanding of the processes involved" and recommended formal exchanges between scientific experts among the A5, improved coordination of government and academic research on Arctic marine ecosystems, and broader international co-operation (NOAA, 2011). Two years later, the United States hosted a meeting of A5 in Washington, DC, where participants reviewed scientific research and recommendations and informally discussed an international agreement to postpone high seas fisheries in the CAO until an appropriate management regime was in place. Following a second FiSCAO meeting in Tromsø, Norway in October 2013, representatives from the A5 met to negotiate fisheries governance issues in Nuuk, Greenland in February 2014. Although participants concurred that commercial fishing in the high seas area of the CAO was unlikely to occur in the near future, and that the establishment of an international fisheries management regime (such as a regional fisheries management organization or arrangement) was not needed at that time, they agreed on the desirability of developing appropriate interim measures to deter future unregulated fishing in the area (NOAA, 2013).

The outcome – the A5's non-binding 2015 *Declaration Concerning the Prevention of Unregulated High Seas Fishing in the Central Arctic Ocean* (or *Oslo Declaration*) – committed the Arctic coastal states to refrain from commercial fishing in the CAO until scientific studies assessed the viability and sustainability of fish stocks. The declaration steered away from the controversial notion of an explicit moratorium, instead seeking to "deter unregulated fishing in the future in the high seas portion of the central Arctic Ocean" (A5, 2015). Accordingly, the A5 agreed to

> authorize our vessels to conduct commercial fishing in this high seas area only pursuant to one or more regional or subregional fisheries management organizations or arrangements that are or may be established to manage such fishing in accordance with recognized international standards.
>
> (A5, 2015)

Because there were no fisheries management organizations covering the CAO, the A5 statement was perceived as a *de facto* temporary moratorium that embodied a precautionary approach to the matter, prohibiting commercial fishing until sound scientific data was available to document the fish stocks present in this area (Shepard *et al.*, 2016). The declaration also represented a non-binding agreement, acting more as a signal of intentions than as a solution to the regional fisheries question.

Through the *Oslo Declaration*, the A5 explicitly recognized

> the interest of other States in preventing unregulated high seas fisheries in the central Arctic Ocean and look forward to working with them in a broader process to develop measures consistent with this Declaration that would include commitments by all interested States.
>
> (A5, 2015)

The A5 process had excluded other nations or associations with long-distance fishing capability who had legitimate rights to participate in the future development and management of resources in the CAO. A5 fisheries experts recognized the value of their potential involvement, however, and strongly encouraged broader international participation in solving the issue. "Inclusion of scientists from countries outside the Arctic coastal states would greatly enhance our collective ability to address these priorities and would also likely bring additional resources to support research and monitoring needs," the report of the 2011 Anchorage meeting of fisheries experts highlighted (NOAA, 2011: 10). In this spirit, the five Arctic coastal states invited representatives from scientific institutions in China, the EU, Iceland, Japan, and South Korea to the third FiSCAO meeting in Seattle, Washington (United States) in April 2015 (NOAA, 2015), heralding the new "Arctic 5+5" format.

The A5+5 equation itself underscored that the Arctic coastal states considered themselves the primary stewards of Arctic Ocean fisheries, due to unique geographical positioning, and that the others would play a secondary role (Zou,

2016). This solution reaffirmed the spirit of the *Ilulissat Declaration*, which clearly asserted the primacy of the A5 in Arctic Ocean governance, while acknowledging the reality that the coastal states could not act unilaterally and prescribe activities on the high seas without violating international law. Accordingly, subsequent A5+5 meetings were organized outside of the Arctic Council or any traditional circumpolar forum and held in different Arctic countries.[5]

Over the following two years, delegations from China, the EU, Iceland, Japan, and South Korea echoed the *Oslo Declaration*'s call to prevent unregulated fisheries in the CAO. The A5+5 reached consensus that this region would be subject to one or more regional fisheries management organizations (RFMO/As). Accordingly, discussions focused on the parameters and processes that should be adopted prior to this outcome, accompanied by separate meetings of Scientific Experts on Fish Stocks in the Central Arctic Ocean (FiSCAO). The A5+5 states produced a draft agreement in March 2017 in Reykjavik, Iceland (A5+5, 2017), and reached agreement in principle in Washington, DC that October. A draft agreement was concluded the following month and, after legal and technical review, the final text of the agreement was circulated in the first half of 2018.

On 3 October 2018, the A5+5 officially signed the *Agreement to Prevent Unregulated High Seas Fisheries in the Central Arctic Ocean* (commonly known as the *Central Arctic Ocean Fisheries Agreement* or CAOFA) in Ilulissat, Greenland (A5+5, 2018). The preamble noted that although "commercial fishing is unlikely to become viable in the high seas portion of the central Arctic Ocean in the near future," the agreement was designed "to prevent unregulated fishing in the high seas portion of the central Arctic Ocean through the application of precautionary conservation and management measures as part of a long-term strategy to safeguard healthy marine ecosystems and to ensure the conservation and sustainable use of fish stocks" (article 2). The "Agreement Area" encompassed "the single high seas portion of the central Arctic Ocean that is surrounded by waters within which Canada, the Kingdom of Denmark in respect of Greenland, the Kingdom of Norway, the Russian Federation and the United States of America exercise fisheries jurisdiction." The latter phrase about "exercising" jurisdiction (rather than the usual language of "areas under national jurisdiction" in most fisheries agreements) was crafted to accommodate the politically sensitive issue of the legal status of the Svalbard Fisheries Protection Zone (Schatz *et al.*, 2018/19).[6]

Conclusion

The CAOF Agreement represents an intriguing case study for several reasons. First, as an initiative originally promoted by the A5, it reflects both the possibilities and limitations of conceptualizing the Arctic Ocean as a region dominated by the coastal states. Although the A5 meeting format emerged to downplay the internationalization of the region and to clarify coastal state rights to delimit their extended continental shelves, it was later used as a vehicle to engage with non-Arctic states when the A5 turned their attention to the fisheries issue. The A5 states had to widen the subregional ensemble to accommodate the broader legal realities in light of the rules constitutive of UNCLOS. Including the three Arctic states not part of the A5 (Iceland directly, and Finland and Sweden through the EU) in the deliberations helped to quash criticisms that the A5 was a regional grouping with legitimacy problems. The CAOF Agreement also underscored the fluid institutional basis for Arctic governance, with the A5 adapting to meet the interests of Arctic and non-Arctic states alike. It also exemplified preventive governance, establishing norms, rules, and standards before the fisheries issue became an urgent concern or even a crisis that could complicate regional dynamics and relationships with the rest of the world (Landriault, 2018; Niiranen *et al.*, 2018).

The CAO posed an unforeseen dilemma for Arctic states that commonly asserted their predominant role in Arctic governance. Non-Arctic states' presence in regional forums could be tolerated if they remained in a subordinated position, as observers contributing to scientific research or to deliberations without insisting on substantive decision-making power (see Lackenbauer *et al.*, 2018: 131–52). This stance, predicated on state sovereignty, made sense as long as issues of regional interest centered on areas within national jurisdictions of Arctic states. The CAO constitutes international waters, making non-Arctic states significant stakeholders who have the legal right to undertake commercial fishing without necessarily consulting Arctic states. By extension, Arctic coastal states had to engage meaningful with non-Arctic states, such as China and South Korea, that had the right to refuse regional agreements. On this point, regional governance alone was not enough; global governance actors had to be taken into consideration.

In the end, the CAOF Agreement affirmed the predominant position of the A5, in that the final agreement mirrored the precautionary approach that they had promoted from the beginning of the process. For non-Arctic state signatories, the trade-off was painless given the relatively low stakes involved in CAO fisheries. The agreement signalled that they were constructive actors in regional (and indirectly, global) governance, open to developing rules rather than ruthlessly exploiting Arctic resources. Furthermore, the CAOF Agreement met the interests of both Arctic and non-Arctic states by not imposing an outright ban or moratorium on commercial high seas fishing – only a temporary indefinite abstention from "unregulated" commercial fishing, which can be lifted when certain procedural and substantive conditions are met. The parties retain (with qualifications) their right to authorize commercial fishing by vessels under their flag pursuant to conservation and management measures adopted by existing RFMOs (Schatz *et al.*, 2018). Moreover, non-Arctic states can contribute to efforts to map fish stocks in the ocean, reaffirming how scientific activities can provide an entry into Arctic governance.

Although the international news media presented the CAOF Agreement as an "historic" or a "landmark" agreement (Kramer, 2017; Sevunts, 2018), it is more appropriately considered the start of a political process rather than its culmination. Article 5.1 stipulates that scientific data must inform the start of negotiations for the establishment of RFMOs, although there is no indication about the appropriate time to initiate such a process. The most difficult negotiations are likely to come over the rules and parameters of these RMFOs, notwithstanding the establishment of the overall fishing volume and the distribution of quotas among signatory states. Far from settling these core issues, the CAOF Agreement relegates them to a future date. Hence, the A5+5 assumed a regulatory role in Arctic governance that is likely to reappear in the foreseeable future as experts compile data on fish stocks in the CAO.

Nevertheless, the broader engagement reflected in the CAOF Agreement negotiations sets a precedent for future A5+5 exchanges and international governance on Arctic issues beyond the sovereign purview of the Arctic states. As international law expert Erik Molenaar notes, it represents a landmark agreement in the application of the precautionary approach and an "important step in the gradual transformation of the freedom of the high seas" (Molenaar, 2018). By elevating the Arctic fisheries issue beyond A5 discussions to a more global level, this case study shows the benefits and necessity of co-operating with non-Arctic states to address specific gaps in Arctic Ocean governance. The UN General Assembly's ongoing negotiations towards a global treaty on marine biodiversity beyond national jurisdiction (BBNJ) may provide a welcome framework to coordinate existing instruments, set common environmental standards, and fill regulatory gaps without undermining relevant legal instruments and frameworks that exist at the regional level (De Lucia *et al.*, 2018). This agreement does not relegate the Arctic Council to the sidelines, nor does it encroach on the sovereign jurisdiction of coastal states. Instead, it affirms the intersecting, and often mutual reinforcing, layers of governance at play in the circumpolar Arctic. Regional and global forces can work together to reach mutually beneficial goals.

Notes

1 Whether a coastal state can claim continental shelf beyond 200 nautical miles depends on whether the submarine areas beyond 200 nautical miles are a natural prolongation of its land territory, as determined in accordance with article 76 of UNCLOS.

2 This important consideration is often missing in media narratives that conflate the issues of continental shelves and waters beyond the EEZ by suggesting that claims to extended continental shelves pose a direct threat to freedom of navigation or represent a massive "grab" of living resources by the states involved.

3 It is more likely that any increase in fish stocks would concentrate in areas closer to shore, meaning within areas of national jurisdiction. See, for example, Young (2016: 272); Shepard *et al.* (2016).

4 The 1995 *United Nations Fish Stock Agreement* (FSA), an Implementation Agreement of UNCLOS, addresses straddling and highly migratory fish stocks within international waters and commits states to two important management concepts: the precautionary approach (to err on the side of caution when regulating on fisheries) and ecosystems-based fisheries management (to recognize that changes in one area of an ecosystem can impact other areas). Regional mechanisms to implement the FSA are known as Regional Fisheries Management Organizations (RFMOs) and Arrangements (RFMAs). RFMOs are the heavier institutional mechanisms, including a Secretariat that operates under a governing body of member states that establish who can fish, what they can fish for, and how much they can catch in a certain section of ocean. RFMO management decisions are binding. In comparison, RFMAs are lighter mechanisms that do not possess a Secretariat, can be bilateral in scope, and do not have to be legally binding, but otherwise have in principle the same characteristics as an RFMO. Molenaar and Koivurova (2010: 16–18),

Molenaar and Corell (2009: 6), Food and Agricultural Organization of the United Nations (FAO) (2019). Thanks to Ryan Dean for sharing a draft article on this topic.

5 Washington (December 2015; November 2017), Iqaluit (July 2016), Tórshavn (Faroe Islands, November/December 2016), Reykjavik (March 2017).

6 For a recent controversy between the EU and Norway about access to the local snow crab fishery, see Andreas Osthagen and Andreas Raspotnik, "Crabtacular! Snow crabs on their march from Svalbard to Brussels," 3 April 2018, www.highnorthnews.com/nb/op-ed-crabtacular-snow-crabs-their-march-svalbard-brussels.

References

Arctic-Five (A5). 2015. *Declaration Concerning the Prevention of Unregulated High Seas Fishing in the Central Arctic Ocean.* 16 July. Oslo, Norway: Arctic-Five (A5). www.regjeringen.no/globalassets/departementene/ud/vedlegg/folkerett/declaration-on-arctic-fisheries-16-july-2015.pdf.

Arctic-Five-Plus-Five (A5+5). 2017. *Chairman's Statement, Meeting on the High Seas Fisheries in the Central Arctic Ocean.* 15–18 March. Reykjavik: Arctic-Five-Plus-Five. www.state.gov/e/oes/ocns/opa/rls/269126.htm.

Arctic-Five-Plus-Five (A5+5). 2018. *Agreement to Prevent Unregulated High Seas Fisheries in the Central Arctic Ocean.* 3 October. Brussels, Belgium: European Commission. https://eur-lex.europa.eu/resource.html?uri=cellar:2554f475-6e25-11e8-9483-01aa75ed71a1.0001.02/DOC_2&format=PDF.

Arctic Oceans Conference. 2008. *Ilulissat Declaration.* Ilulissat, Greenland: Arctic Oceans Conference. www.oceanlaw.org/downloads/arctic/Ilulissat_Declaration.pdf

Bailey, Kevin. 2011. "An Empty Donut Hole: The Great Collapse of a North American Fishery." *Ecology and Society* 16, no. 2: 28–41.

Baker, Betsy. 2009. "Filling an Arctic Gap: Legal and Regulatory Possibilities for Canadian-US Cooperation in the Beaufort Sea." *Vermont Law Review* 34: 57–120.

Baker, Betsy and Brooks Yeager. 2015. "Coordinated Ocean Stewardship in the Arctic: Needs, Challenges and Possible Models for an Arctic Ocean Coordinating Agreement." *Transnational Environment Law* 4, no. 2: 359–394.

Borgerson, Scott G. 2008. "Arctic Meltdown: The Economic and Security Implications of Global Warming." *Foreign Affairs* 87: 63.

Boswell, Randy. 2010. "Arctic Native Leaders Say They've Been Left Out of Summit." *Canwest News Service*, February 15.

Canadian Broadcasting Corporation (CBC) News. 2010. "Clinton's Arctic Comments Cheer Inuit." March 31.

Cannon, Lawrence. 2010. "Arctic Ocean Coastal States Meeting, Chelsea, March 29, 2010, Summary by Lawrence Cannon, Foreign Affairs Minister of Canada." www.international.gc.ca/polar-polaire/arctic-meeting_reunion-arctique-2010_summary_sommaire.aspx?lang=en&view=d

Commission on the Limits of the Continental Shelf. 2018. "Submissions and Recommendations." www.un.org/Depts/los/clcs_new/commission_submissions.htm

De Lucia, Vito, Christian Prip, Kristine Dalaker Kraabel, and Raul Primicerio. 2018. "Arctic Marine Biodiversity in the High Seas between Regional and Global Governance." *Arctic Review* 9: 264–266.

Elferink, Alex Oude and Constance Johnson. 2006. "Outer Limits of the Continental Shelf and Disputed Areas: State Practice concerning Article 76 (10) of the LOS Convention." *International Journal of Marine & Coastal Law* 21, no. 4: 461.

Food and Agricultural Organization of the United Nations (FAO). 2019. "Regional Fisheries (RFB)." www.fao.org/fishery/rfb/en. Last accessed June 14, 2019
Global Affairs Canada. 2016. "Defining the Extended Continental Shelf." www.international.gc.ca/arctic-arctique/continental/limits-continental-limites.aspx?lang=eng. Last accessed August 25, 2016.
Government of Canada. 2016. "United States-Canada Joint Arctic Leaders' Statement." December 20. https://pm.gc.ca/eng/news/2016/12/20/united-states-canada-joint-arctic-leaders-statement.
Hertell, Hans H. 2009. "Arctic Melt: The Tipping Point for an Arctic Treaty." *Georgetown International Environmental Law Review* 21: 565–591.
Hollowed, Anne, Benjamin Planque, and Harald Lg. 2013. "Potential Movement of Fish and Shellfish Stocks from the Sub-Arctic to the Arctic Ocean." *Fisheries Oceanography* 22, no. 5: 355–370.
Huebert, Robert. 2009. "The Need for an Arctic Treaty: Growing from the United Nations Convention on the Law of the Sea." *Ocean Yearbook* 23: 27–37.
Huebert, Robert and Brooks B. Yaeger. 2008. *A New Sea: The Need for a Regional Agreement on Management and Conservation of the Arctic Marine Environment*. Oslo, Norway: WWF International Arctic Programme.
Jares, Vladimir. 2009. "The Continental Shelf Beyond 200 Nautical Miles: The Work of the Commission on the Limits of the Continental Shelf and the Arctic." *Vanderbilt Journal of Transnational Law* 42: 1265–1305.
Koivurova, Timo and Erik J. Molenaar. 2009. *International Governance and Regulation of the Marine Arctic: Overview and Gap Analysis*. Oslo, Norway: World Wildlife Foundation, 2009.
Kramer, Andrew. 2017. "Russia, U.S. and Other Nations Restrict Fishing in Thawing Arctic." *New York Times*, November 30. www.nytimes.com/2017/11/30/world/europe/russia-arctic-ocean-fishing-thaw.html.
Lackenbauer, P. Whitney, Adam Lajeunesse, James Manicom, and Frédéric Lasserre. 2018. *China's Arctic Ambitions and What They Mean for Canada*. Calgary, Alberta: University of Calgary Press.
Landriault, Mathieu. 2018. "Opening a New Ocean: Arctic Ocean Fisheries Regime as a (Potential) Turning Point for Canada's Arctic Policy." *International Journal* 73, no. 1: 158–165.
Lavrov, Sergei and Jonas Gahr Støre. 2010. "Canada, take note: Here's how to resolve maritime disputes." *Globe and Mail*, September 21. www.theglobeandmail.com/opinion/canada-take-note-heres-how-to-resolve-maritime-disputes/article4326372/.
Livermore, Ian. 2017. "From Cold War to North Pole Alliance: Canada and the Changing Face of Defence and Foreign Relations in the Arctic." In *Whole of Government through an Arctic Lens*, edited by Whitney Lackenbauer and Heather Nicol, 52–113. Antigonish, Nova Scotia: Mulroney Institute of Government.
McDorman, Ted. 2010. "The Outer Continental Shelf in the Arctic Ocean: Legal Framework and Recent Developments." In *Law, Technology and Science for Oceans in Globalisation*, ed. Davor Vidas, 499–520. Leiden: Martinus Nijhoff.
Molenaar, Erik J. 2012. "Arctic Fisheries and International Law: Gaps and Options to Address Them." *Carbon & Climate Law Review* 6, no. 1: 63–77.
Molenaar, Erik J. 2018. The CAOF Agreement: Key Issues of International Fisheries Law." Presentation to the conference on *New Knowledge and Changing Circumstances in the Law of the Sea*. 28–30 June. Reykjavik, Iceland. http://icelandkmiconference2018.com/wp-content/uploads/2018/07/Molenaar-presentation-CAOF-Agreement.pdf.
Molenaar, Erik J. and Robert Corell. 2009. "Arctic Fisheries." *Arctic TRANSFORM Project*. http://arctic-transform.org/download/FishBP.pdf.

Molenaar, Erik J. and Timo Koivurova. 2010. *International Governance and Regulation of the Marine Arctic*. Copenhagen: WWF International Arctic Programme. Last accessed June 14, 2019. http://awsassets.panda.org/downloads/3in1_final.pdf.

Niiranen, Susa, Andreas Richter, Thorsten Blenckner, Leif Christian Stige, Matilda Valman, and Anne-Maria Eikeset. 2018. "Global Connectivity and Cross-Scale Interactions Create Uncertainty for Blue Growth of Arctic Fisheries." *Marine Policy* 87: 321–330.

NOAA. 2009. "Fish Resources of the Arctic Management Plan." www.fisheries.noaa.gov/management-plan/fish-resources-arctic-management-plan.

NOAA. 2011. *Report of a Meeting of Scientific Experts on Fish Stocks in the Arctic Ocean Anchorage, Alaska, June 15–17, 2011*. Anchorage, Alaska: NOAA. www.afsc.noaa.gov/Arctic_fish_stocks_ third_meeting/First%20Meeting%20Sci%20Experts%20Arctic%20Fisheries%2030%20Aug%202011.pdf.

NOAA. 2013. *Report of the 2nd Scientific Meeting on Arctic Fish Stocks, Tromsø, 28–31 October 2013*. Tromsø, Norway: NOAA. www.afsc.noaa.gov/Arctic_fish_stocks_third_meeting/Report%20of%202nd%20Scientific%20Meeting%20on%20Arctic%20Fish%20Stocks%2028%2031%20October%202013.pdf.

NOAA. 2015. "Third Meeting of Scientific Experts on Fish Stocks in the Central Arctic Ocean." www.afsc.noaa.gov/Arctic_fish_stocks_third_meeting/.

Reichert, Christian. 2009. "Determination of the Outer Continental Shelf Limits and the Role of the Commission on the Limits of the Continental Shelf." *International Journal of Marine & Coastal Law* 24: 387–399.

Riddell-Dixon, Elizabeth. 2008. "Canada and Arctic Politics: The Continental Shelf Extension." *Ocean Development and International Law* 39, no. 4: 343–359.

Schatz, Valentin, Alexander Proelss, and Nengye Liu. 2018. "The 2018 Agreement to Prevent Unregulated High Seas Fisheries in the Central Arctic Ocean: A Primer." *EJIL Talk*, October 26. www.ejiltalk.org/the-2018-agreement-to-prevent-unregulated-high-seas-fisheries-in-the-central-arctic-ocean-a-primer/.

Schatz, Valentin, Alexander Proelss, and Nengye Liu. 2019. "The Agreement to Prevent Unregulated High Seas Fisheries in the Central Arctic Ocean: A Critical Analysis." *International Journal of Marine and Coastal Law* 34: 195–244.

Sevunts, Levon. 2018. "Canada, EU and 8 Other Countries Set to Sign 'Historic' Agreement to Protect the Central Arctic Ocean." *CBC News*. www.cbc.ca/news/politics/canada-treaty-protect-arctic-ocean-1.4847890.

Shepard, Grace, Kari Dalen, Regina Peldszus, Sara Aparicio, Larissa Beumer, Roger Birkeland, Nikolaos Gkikas, Melina Kourantidou, Panagiotis Ktenas, Peter Wilhelm Linde, Francesco Marazzi, Ruth Pincinato, Jaziar Radianti, Bernhard Schartmuller, Eike Ingrid Stubner, Alix Varnajot, Magali Vullierme, and Irina Zhilina. 2016. "Assessing the Added value of the Recent Declaration on Unregulated Fishing for Sustainable Governance of the Central Arctic Ocean." *Marine Policy* 66: 50–57.

Solli, Per Erick, Elana Wilson Rowe, and Wrenn Yennie Lindgren. 2013. "Coming into the Cold: Asia's Arctic Interests." *Polar Geography* 36, no. 4: 253–270.

Young, Oran. 2009. Whither the Arctic? Conflict or Co-operation in the Circumpolar North." *Polar Record* 45, no. 1 (2009): 73–82.

Young, Oran. 2016. "Governing the Arctic Ocean." *Marine Policy* 72: 272

Zou, Leilei. 2016. "Stirred Waters Under the Ice Cap: An Analysis on A5's Stewardship in the Central Arctic Ocean Fisheries Management." *Arctic Yearbook* 5: 411–424.

Zou, Leilei and Henry P. Huntington. 2018. "Implications of the Convention on the Conservation and Management of Pollock Resources in the Central Bering Sea for the Management of Fisheries in the Central Arctic Ocean." *Marine Policy* 88: 132–138.

3 From international to regional

Sub-national units in Arctic governance

Traditionally, political analysts and international relations scholars alike have ignored or downplayed the role played by sub-national governments at the international level. This is understandable given that sovereign states exercise jurisdiction over military matters, treaty-making, and (most) diplomatic representation. World affairs, however, do not focus solely on hard power or high politics, and have diversified since the 1970s to include a wide range of so-called "low-politics" issues – such as environmental protection, education, health, infrastructure, economic development, natural resources, and social services – which are subjected to pressures beyond national borders. Regional or local governments administer or deal with many of these issues, making sub-national units core stakeholders in global and regional governance. International action carried out by sub-national units is known as paradiplomacy.

In the Arctic region, the phenomenon of paradiplomacy is even more salient; as Chapter 1 explains, Arctic states have been active in co-operating on environmental and social issues, as well as on environmental protection. This chapter shows how and why sub-national governments can be understood as complementary actors to sovereign states: regional governments are now active contributors to Arctic governance through diplomatic efforts supporting the creation of norms and standards. The highly fragmented configuration of the Arctic region, differences between national political systems, and the localized nature of competencies by these political actors impede more integrated action from these Arctic sub-national units.

Paradiplomacy

Sub-national entities, and particularly regional governments, are active players in the international realm, "negotiating and signing international agreements, developing representations abroad, conducting trade missions seeking foreign investment, and entering into bilateral and multilateral relations with states" (Lecours, 2002). Although regional governments partake in activities traditionally associated with sovereign states, different dynamics and pressures orient their behaviors. For example, sub-national governments work in an environment riddled with constraints; they are typically dependent upon their national government for the amount of freedom they possess on the international stage. Traditionally, this has led sub-national governments to pursue pragmatic international endeavours in order to avoid jurisdictional confrontation or conflict with their national government. Accordingly, sub-national actors prefer problem-solving local issues to grandstanding or confrontational politics (Joenniemi and Sergunin, 2014). Keating (2000) argues that sub-national units "cannot simply lay down a line to be followed by all but must seek to bring together independent actors around specific programmations and issues." He astutely observes that "they must fit their own activities into a world dominated by national governments and transnational organizations, which they can rarely challenge head on but must work around or with."

Sub-national actors tend to direct their international activity towards areas of exclusive competencies (especially in federal states) and on functional co-operative endeavours in areas of low politics. Regional governments may also act as a multiplier on the trade and investments front, adding to efforts made by their central government to attract investments as well as to generate trade and economic growth. As such, regional governments complement national governments by pursuing economic, social, and cultural goals (McHugh, 2015: 239). These types of initiatives can be considered as part of a pragmatic strategy incorporated under the functional paradiplomacy label.

Regions in which a strong nation-building project and/or self-determination movement is in place (such as Québec or Scotland) are, of course, not only moved by functional reasons; external projection becomes a tool to display a distinct national identity, often in collaboration with other social and economic actors (Keating, 2013). Similarly, the level of co-operation with the central government fluctuates depending on the level of ambition displayed by the nationalist movement. In these cases, we are confronted with identity paradiplomacy: sub-national actors tend to promote the distinctiveness of their values and interests. The desires to achieve further recognition and legitimacy as a nation motivate the international policy of these regional governments (Paquin, 2004; Massie and Lamontagne, 2018). Identity paradiplomacy can rapidly turn into protodiplomacy (the effort by a sub-national government to be active in competencies exclusive to the central government), which is typically implemented when regional governments "promote claims of political independence or autonomy" (McHugh, 2015).

Notwithstanding the nature of their international actions, the potential contribution of sub-national entities to Arctic governance is obvious. In federal states, these governments can pass legislation and invest substantial funds while sub-national entities in unitary states administer local services and implement programs. In both cases, legislations and service administration embody specific standards and norms. Sub-national entities also add a layer of representativeness: they are political units closer to their constituents and govern a more limited political community than national governments.

Accordingly, Arctic paradiplomacy is characterized by dominant functional patterns, with a pragmatic focus on local issues such as regional economic development, infrastructure, natural resources exploitation, and environmental protection. Sub-national governments also tend to reinforce the fragmentation of the Arctic region, creating sub-regions (North America, Euro-Barents) constituted of collaborative networks and co-operative endeavours. Identity paradiplomacy has come forward through forums such as the Arctic Circle Assembly, however, with actors such as Québec and Scotland incorporating the Arctic within their broader international policy.

Arctic paradiplomacy – governance between sub-national units

The nature of political systems found in the Arctic region varies considerably, with three federal states and five unitary national governments. This situation, in turn, allows for an interesting sample of possible paradiplomatic activities, even if it limits the potential of co-operation between these units (see Table 3.1.).

There is substantial disparity in the levels of autonomy granted to sub-national entities in Arctic states. Federal states such as Canada and the United States feature considerable decentralization towards provinces/territories and states, while the Russian government displays more centralist tendencies while according different levels of independence to sub-national units (Dubreuil, 2011). On the other hand, the unitary state of Denmark delegates important competencies to Greenland and the Faroe Islands. Thus, competency areas vary significantly from one regional government to another.

Table 3.1 Type of political system in Arctic states.

	Type of system	*Sub-units*	*North of the 60th parallel*
Russia	Federalism	Republics, Oblasts, regions, and districts	Karelia, Komi, Kamchatka and Sakha, among others
Canada	Federalism	Provinces and territories	Nunavut, Yukon, Northwest Territories, Québec
United States	Federalism	States	Alaska
Iceland	Unitary	Provinces	Westfjords, Northeastern, among others
Norway	Unitary	Counties	Finnmark, Svalbard, Troms, among others
Sweden	Unitary	County council	Norrbotten, Västerbotten, among others
Finland	Unitary	Regions	Lapland, Northern Ostrobothnia, among others
Denmark	Unitary	Self-governing units, state administrations	Greenland, Faroe Islands

Source: With information from Anderson (2008) and Dubreuil (2011).

The precursor: the Northern Forum

The creation of the Northern Forum (NF) in 1991 represented the first attempt at Arctic paradiplomacy. "By 1990, the world had changed and the opening of the USSR to the outer world provided northern Russian regions with the opportunity to join in the creation of a 'voice of northern regions,'" Hasanat notes (2012a: 373). This co-operative turning point prompted 19 sub-national units from nine different countries (including the Soviet Union) to collaborate and agree in principle to the creation of the NF.

The Forum's composition is unlike that of any other Arctic co-operative initiatives of the time (for example, the Arctic Environmental Protection Strategy, or AEPS), as only sub-national governments could become members. Criteria for membership differed from other Arctic forums because the conditions for eligibility were defined in broad terms. Regional governments are described as "a governmental entity[1] from a region meeting basic characteristics of a northern region or a region interested in involvement in international co-operation in the Arctic and the North" (Northern Forum, 2013). The basic characteristics associated with northern regions raise a few questions of interpretation, as the 2013 revised rules of procedure do not explicitly state them. However, past practices recognized broad demographic, climatic, economic, and political criteria.

Criteria for recognition as a northern region in the Northern Forum

- Harsh climate and vulnerable ecosystems;
- Small population, diverse and strong indigenous cultures;
- Economy based primarily on natural resources extraction and outsourcing;
- Limited internal investment capital;
- Limited infrastructure, oriented North-South;
- Limited influence on national government

Listed in Hasanat (2012a).

The NF is perceived as a soft-law institution with low institutional development (Young, 2002, 2005; Hasanat, 2012a). For Young (2002), the NF plays a catalyst role in bringing people together to discuss matters of common interests; it also assumes a generative role by promoting new understandings and disseminating new perceptions about northern regions to the world. On the other hand, it has historically displayed a low potential for assuming a programmatic role that would be action- or investment-oriented.

Throughout its existence, the NF has promoted norms and standards through declarations and statements, with limited success and diffusion (Hasanat, 2012a). The common environmental domain norm is probably its most important contribution, and represents a precursor of norms developed by Indigenous groups afterwards. As articulated in the 1994 *Rovaniemi Declaration*, relations between people and nature are based on essential co-existence. Hence, the objective is to avoid damaging environmental integrity and to respect the rights of the local population to a clean environment (Hasanat, 2012b: 279). In itself, the concept of common environmental domain reasserts the notion of sustainable development as described in previous statements by international organizations in the late 1980s until the early 1990s.

In this case, the promotion of sustainable development norms undermined the popular understanding that northern regions were solely areas of pristine nature. For example, the Forum's 23rd resolution, adopted on 16 September 1994, called on the AEPS to work to "remove barriers to trade in the use of our resources, including animals." In the same vein, the NF emphasized building healthy economies, the exploration and development of non-renewable resources, and the development of efficient production systems (Northern Forum, 1994). Economic development is an intricate part of the normative promotion of the NF, which presents northern regions as areas requiring more economic activity subject to the consent of local populations.

As an observer to the Arctic Council (AC), the NF continued to advocate for this priority. In their 2014–2016 report, the organization listed sustainable development goals, Indigenous rights, and energy development as areas of common interest with the AC. The first priority emphasized a perceived imperative to "evaluate the economic and living conditions of Arctic peoples, identify barriers to socio-economic development and produce a best practice guide that highlights models for improving the region's quality of life" (Pogodaev, 2016). Overall, six of the NF's nine priority directions relate closely to socio-economic needs and economic development; these include the development of small and medium businesses, the implementation of business potential for youth, and the development of transport and telecommunication infrastructure in the North, among other priorities (Pogodaev, 2016).

The NF is also active in developing common projects by pooling resources and achieving common objectives. Contrary to vague or well-intended statements, practices entail substantial investments from sub-national entities (and not just words). The priority projects implemented illustrate the NF's inherent limitations and its limited potential to generate norms by committing to actions or investments. Two observations can be deducted when analyzing NF projects for the 2015–2018 period (Table 3.2). First, only the "Red" book project can be considered a true regional project involving a significant number of NF members. Half of the other projects involved only one or two NF regional members, and hinted at geographically circumscribed regional co-operative ties. Two regional governments (Khanty-Mansiysk and Sakha) are spearheading several NF activities, while others (Akureyri, Lapland, and Gangwon) are inactive in most NF priority projects.

Second, the organization has not evolved significantly since the early 2000s. NF projects are still predominantly filling catalyst and generative roles. For example, the northern zoos, Youth Environmental Forum, and specially protected areas projects are about bringing interested parties together to exchange experience and share best practices. On the other hand, the water and climate change, "Human in

Table 3.2 List of Northern Forum priority projects (2015–2018).

Projects	*Nature of initiative*	*NF participants*	*Role*
Prevention of Addictions	Health	2 (Khanty, Sakha)	Programmatic
Suicide Prevention	Health	2 (Khanty, Sakha)	Programmatic
Infectious Diseases Control	Health	4 (Khanty, Sakha, Chukotka, Kamchatka)	Generative
Telemedicine	Health	4 (Khanty, Sakha, Alaska, Gangwon)	Programmatic
Specially Protected Areas	Environment	2 (Sakha, Krasnoyarsk)	Catalyst
Schools Partnership	Education	3 (Sakha, Krasnoyarsk, Alaska)	Catalyst
Festival of Northern Fishing Traditions	Culture	2 (Sakha, Lapland)	Generative/ catalyst
Red Book	Education	11(All Russian members and Alaska)	Generative
International Arctic Center for Culture and Arts	Culture	1 (Sakha)	Generative
Human in the Arctic	Economic	1 (Sakha)	Generative
Energy Supply for Isolated Regions	Economic	1 (Sakha)	Catalyst/ Programmatic
Brown Bear	Environmental	4 (Chukotka, Alaska, Khamchatka, Khanti)	Generative/ catalyst
Youth Environmental Forum	Environmental	3 (Sakha, Khanty, Yamal)	Catalyst
Water and Climate Change	Environmental	4 (Khanty, Alaska, Sakha, Krasnoyark)	Generative
Northern Zoos	Environmental	5 (Sakha, Alaska, Lapland, Krasnoyark, Khabarovsk)	Generative/ catalyst
International Arctic School	Education	1 (Sakha)	Programmatic

Source: Information taken from Northern Forum (2018).

the Arctic," and infectious disease control projects are designed to create knowledge and to gather evidence to better address these problems.

Only five NF projects invested in actions to address a specific issue. These programmatic activities tended to have fewer members involved, raising the question as to whether these initiatives could have been implemented by these regional governments without NF support. The telemedicine project is probably the most interesting co-operative venture, involving technology transfer and the exchange of expertise to establish training programs. It facilitated access to medical care remotely in two Russian regions through practical assistance from Alaska and Gangwon. The health sector more generally has yielded the most concrete results by developing indicators, sharing technology and adapting health care practices to

Northerners' specific medical needs, especially alcoholism and tuberculosis. Close co-operation with the scientific community has proven crucial for these projects to come to fruition. Projects of a catalyst nature can also produce tangible results. The energy supply for isolated regions project had a primary objective to bring together civil society organizations, businesses, and regional governments and have them share views on the issue of renewable energy resources, leading to investments in the renewable energy industry (Northern Forum, 2018). However, the outcomes of projects of a catalyst nature are uncertain; bringing stakeholders together does not mean change in policies or investments.

The innovative structure of the NF reinforced a norm of including non-state actors in Arctic governance mechanisms. Through a business partnership category (Membership II status), corporations gain an official role in the NF that allows them to attend all Forum meetings and participate in projects (Northern Forum, 2013), reinforcing the desirability of public-private partnerships. Businesses see the NF as an opportunity to expand their market reach and create ties with sub-national governments and other businesses. At the same time, this membership category meets one of the NF's core mandates to improve the quality of life of Northerners, invest in infrastructure, and promote sustainable development (particularly in the resource extraction sector) (Young, 2005; Hasanat, 2012a).

Despite the positive initiatives, fundamental structural shortcomings help to explain the NF's low impact on Arctic governance more broadly. According to Hasanat (2012a), the Forum promoted the concept of northern regions in an innovative way, making a significant contribution to international law. In our view, the extreme broadness of the membership criteria renders the concept inoperable. The limited autonomy criterion applied for Canadian provinces (such as Alberta, for example), counties in Norway, provinces in China, and municipalities in Iceland – each with very different autonomy vis-à-vis central authorities. Similarly, the territory of Nunavut (Canada), the Yamal-Nenets Autonomous Oblat (Russia), the Hokkaido prefecture (Japan), and the Gangwon Province (South Korea) all meet the harsh climate criterion, even though these regional governments serve territories with significant climate variations, which blurs the line between northern regions and peripheral areas.

The NF's loose membership criteria limits its role in Arctic governance. Contrary to other forums or soft-law arrangements, the organization has an incredibly high rate of member turnover. Out of the 12 founding members, only 6 were still members of the forum as of 2018; the NF reached a peak of 25 members from 2001 to 2003 and a low of 7 members from 2013 to 2014 (Tsui, 2016). As of 2018, 10 out of the 14 members are from Russia, increasing coherence and common purpose but limiting influence in the Arctic region. At the same time, members have little shared infrastructure and few possibilities for cross-boundary co-operation, as the projects implemented from 2015–2018 indicate. As Russian sub-national governments turn their attention to the international stage to promote sustainable development, improve their international attractiveness, and carve themselves a sphere of relative autonomy vis-à-vis Moscow (Joenniemi and Sergunin, 2014), they find little opportunity in a Northern Forum dominated by fellow Russian regions.

Consequently, many regional governments have withdrawn from the NF to pursue co-operative ventures in other forums. Regional governments have limited resources, making them selective in how they pursue international ties. For example, the government of Québec joined the NF in 2010 but became inactive two years later and withdrew officially in 2013, citing their participation at the Arctic Circle Assemblies and Forums to justify their non-participation (Northern Forum, 2018). Limited resources make active participation in all forums pertaining to Northern affairs impossible.

In other cases, regional governments have reoriented their efforts to regional settings, allowing greater shared interests between members. In the Nordic countries, Troms and Norbotten withdrew from the NF to concentrate their resources on the Barents Regional Council, part of the co-operation fostered in the Barents Euro-Arctic Region (BEAR). Northwestern Russian regions have remained members of the NF, but are also more active in BEAR co-operation. This shift illustrates the emphasis on functional paradiplomacy to address local issues that transcend national boundaries.

Transboundary issues imply high geographical coherence in order to enhance pragmatic, co-operative ties. Generative functions (generating and diffusing new perceptions about the Arctic region abroad) might be a worthwhile endeavour in some contexts and for some actors, but regional governments tend to invest their limited resources in more pragmatic and productive initiatives. Cross-border co-operation becomes a natural priority for these governments.

The Barents Euro-Arctic Region: functional paradiplomacy at play

The Barents Sea region encompasses the northernmost regions of Norway, Sweden, Finland, and Northwestern Russia. These four countries share thousands of kilometers of common borders, with millions of people crossing every year at the Finland-Russia border alone, giving rise to a range of transport, economic, and environmental issues. Furthermore, three Indigenous peoples call the region their home: the Nenets, the Vepsians, and the Saami (with the latter dispersed in all four countries). Moreover, these regions are all dependent on their national government along a north-south axis. Challenges faced by the different regional authorities are strikingly similar, including high unemployment rates, an ageing population, and lacking transportation infrastructures (BEAR, 2014).

Accordingly, the Barents Sea region has the potential to supplement the predominant north-south orientation within the nation states with pragmatic co-operation following an east-west axis across national borders. For example, economic co-operation between Finnish and Russian regions was already well underway in the 1980s and 1990s, with joint projects related to bus service, water treatment plan and natural resources transformation (Shevchuk, 2014). Accordingly, Barents Euro-Arctic Region (BEAR) co-operation resulted from the new post-Cold War strategic environment and the associated peace dividend. In 1993, the European Commission, Russia, and Nordic countries (Finland, Norway, Sweden) signed the

Kirkenes Declaration to enhance regional stability, progress, and international peace and security. Co-operation was not formalized through an international treaty, however, which produced uncertain institutionalization and weak regional symbols (Myrjord, 2003; Hasanat, 2010; Zimmerbauer, 2013; Bailes, 2013).

BEAR co-operation did, however, create institutions intended to foster regional ties. Norway spearheaded the initiative, with the primary objective of constructively engaging with post-communist Russia through multilateral channels (Leland and Hoel, 2008). Sub-national entities in the Barents Sea region promoted and facilitated the initiative, conceiving of BEAR as a two-layered structure involving both national and sub-national governments (Zimmerbauer, 2013). On the first level, the Barents Euro-Arctic Council (BEAC) is composed of sovereign states and one supranational entity, with Denmark, Finland, Iceland, Norway, Russia, Sweden, and the European Commission as members. On the second level, regional governments co-operate in the Barents Regional Council (BRC), encompassing sub-national authorities from Finland, Norway, Russia (Northwestern regions), and Sweden.[2] In contrast to the Northern Forum (NF), the BEAR has limited membership geographically, with the current membership representing the maximal extent of the regional co-operation project. Members such as Norway expressed particular reluctance to include more members, fearing that such a move would jeopardize "the Arctic character of the region" (Zimmerbauer, 2013: 99).

As part of the *Kirkenes Declaration*, governments agreed to co-operate on several different fronts: environmental issues; economic, scientific, and technological co-operation; regional infrastructure; Indigenous peoples' issues; cultural relations; and tourism (BEAR, 1993). These priorities reflect the interests of four working groups created in BEAC, the three BRC working groups, and six joint working groups. The organization has had a significant impact on communicable disease control, pollution, and conservation, in terms of both actions and standards to influence best practices. Moreover, the councils have sustained initiatives on those two fronts, adopting a long-term perspective (Stokke and Hønneland, 2007).

The close co-operation between the BEAC and BRC is unsurprising considering that all national governments are unitary (with the exception of Russia, which displays strong centralizing pressures). Hence, the contribution of the sub-national entities must be weighted in terms of their participation in the BRC working groups and in the joint working groups. Regional governments also co-operate with other regional bodies, such as the Nordic Council of Ministers and the Arctic Council.

The impact of regional governments on norms and standards must be analyzed in light of the projects they have implemented. Member countries have heavily invested in bilateral co-operative ties, mostly between Russia and either Norway or Finland (Zimmerbauer, 2013). Norway alone invests more than $4 million US annually in co-operation projects with Russia through its Norwegian Barents Secretariat. A total of 4,000 projects have been financed since 1993 through this mechanism, many by sub-national entities (Norwegian Barents Secretariat, 2018).

An analysis of multilateral initiatives spearheaded through the Barents co-operation institutions reveals contrasts with the NF (listed in Table 3.3). First, projects typically involve multiple regional partners, highlighting the concerted effort and

Table 3.3 List of projects under the framework of the BEAR co-operation since 2014.

Project	*Year*	*Nature of initiative*	*Members*	*Role*	*Russia as recipient*
Pasvik Monitoring Program	2003–2014	Environmental	3 (Finnmark, Lapland, Murmansk)	Generative	Yes
Bioeconomy in the North – Forest and waste management	2017–2019	Environment	6 (Nordic States, Arkhangelsk, Karelia, Murmansk)	Programmatic	Yes, sole recipient
B2B Business to Barents	2018–2019	Economy	14 (led by North Karelia)	Catalyst	Yes
Kolarctic CBC programme	2014–2020	Transport	13 (led by Kainuu)	Generative	Yes
Karelia CBC Programme	2014–2020	Environmental/ Economy	4 (Kainuu, North Karelia, Oulu, Karelia)	All three	Yes
Finnish-Russian bilateral project AB-Waste	2017–2018	Environmental	9 (Finnish and Russian regions)	Catalyst	Yes
Telemedicine	2017–2020	Health	2 (Troms, Nenets)	Programmatic	Yes, sole recipient
Programme on HIV/AIDS	2015–2019	Health	14 (all members)	Programmatic	Yes
The Barents Tuberculosis Programme	2016–2019	Health	14 (all members)	Programmatic	Yes
Children and Youth at Risk	2016–2019	Health	14 (all members)	Programmatic	Yes

coordination among regional governments. Second, BRC members have pursued generative and catalyst functions, which is typical of sub-national actors. Many projects have also played a programmatic role by pooling resources together and coordinating efforts to achieve a common goal.

Regional governments actively co-operate in BRC working groups on the environment, transport and logistics, and investments and economic co-operation. Environmental co-operation tends to focus on reduction of pollutants rather than climate change, and numerous initiatives have been implemented under the general guidelines of the subgroup on Hot Spots Exclusion (SHE). The BEAC lists 42 sites posing significant ecological risks, all of which are located in Northwest Russia. The objective is to remediate these sites so that they no longer constitute hazards to nature and communities. The Pasvik Monitoring Program and the Bioeconomy in the North projects constitute part of this collective effort: 12 sites have been cleaned, with six more in the assessment stage as of fall 2018 (BEAC, 2018b). The promotion of standards finds a concrete application in these cases. For example, the transformation of wood waste generated by a timber processing industry in Komi into biofuel required technological changes and capacity building through training and advice from Norwegian counterparts. This resulted in the removal of this site from the SHE list, creating jobs, reducing the costs of fuel, and limiting environmental impacts (BEAC, 2017).

Transportation and health issues also led to thick co-operative networks. Relevant health projects are directly aimed at tackling diseases ravaging Northern communities, especially tuberculosis and HIV/AIDS, with norms of prevention and harms-reduction as central objectives. Telemedicine projects, on the other hand, address Northern communities' difficulties accessing medical care, with a new Barents programme (Rural Health for Barents) which directs regional participants to e-health projects and distance medical care as a way to deal with northern isolation (BEAR, 2018).

Pragmatic and active co-operation between BEAR members grapples with transportation problems to "create conditions for cross-border passenger transport" by conceptualizing the Barents region as a single transport area (BEAC, 2018a). The development of an integrated transport infrastructure, as laid down in the Joint Barents Transport Plan, reflects similarities across the region: low traffic flows, reliance on natural resources extraction, difficult wintertime maintenance, and isolated communities (BEAR, 2015). Given the inherent linkages between transportation, economic, and environmental issues, optimized transportation promises to contribute to "sustainable development of businesses, higher employment rates and the quality of life in all territories of the Northern Periphery of Europe" (International Barents Secretariat, 2018). Co-operation in recent years has followed similar patterns observed by Zimmerbauer in 2013: economic and business collaboration has intensified, with special attention paid to pollution reduction and waste-management initiatives.

Barents collaboration, especially on health and environmental issues, is marked by a strong east-west orientation, with Northwest Russia as the primary or sole recipient and the Nordic countries serving as aid providers. For example, Nordic governments provide Russian governments with technology, advisory assistance, and expertise to implement bio-economic industries. The adoption and implementation of these industrial and environmental standards produce savings and technological advancement on the Russian side, while reducing the risk of environmental hazards on the Nordic side.

As illustrated above, functional paradiplomacy represents the dominant principle guiding sub-national actors' behaviors and decisions in the BEAR. Economic, environmental, and transportation issues embody low politics and pragmatic co-operation, with BEAR institutions allowing regional governments to co-operate closely with national governments. The hybrid structure facilitates co-operative endeavours between both levels through joint working groups, with BEAC and BRC evolving concurrently as connected rather than distinct entities. Regional governments influence national governments via work and assessments conducted at the BRC, while national governments retain the upper hand and offer the general BEAC orientations. Funding is also decided at the national level, and influences the action of sub-national governments. The inclusion of other partners, such as businesses and non-governmental organizations, also highlights how sub-national governments conceptualize governance as an inclusive project.

The North American Arctic

While BEAR cooperation has produced positive results, other sub-regions in the circumpolar North have not developed such institutional capacity. The potential to bring together sub-national entities in the North American Arctic exists. Alaska, Greenland, and the three Canadian territories (Northwest Territories, Yukon, and Nunavut) share a heavy reliance on natural resources extraction, dire socio-economic needs, isolated communities, and Indigenous cultures. They also are overwhelmingly impacted by global warming and share an interconnected space, often with common boundaries. These similarities, and a dramatic moment of opportunity and worry with the end of the Cold War, led to the thick co-operative ties formed in the Barents Euro-Arctic Region. Yet, no forum or association was formed to facilitate co-operation between these North American governments.

Reinforcing these trends is the fact that many sub-national governments routinely prioritize pressuring and negotiating with their central government for additional powers. Canadian territories have been looking for years to the Government of Canada to obtain complete jurisdiction over natural resources royalties, as well as control over lands and resources. The Government of Canada put in place a cap on revenues on resource development that the Northwest Territories and Yukon could keep (Speca, 2012). Further, Canadian territories already coordinate their efforts and collaborate through mechanisms such as the Northern Premiers' Forum, which have allowed leaders of the three Canadian territories to meet annually since 2002. However, this type of forum tends to emphasize collaborative ties within Canada rather than a truly North American Arctic perspective.

Greenland, for its part, still displays a hesitant international posture. Increased autonomy was obtained in 2009, with considerable control over natural resources revenues, culture, and education. Greenland could also "participate in negotiations for agreements that concern it and may conclude agreements with foreign states and international organizations that exclusively concern Greenland" (Auchet, 2011: 962). However, the territory must obtain the approval of Denmark to make the process official. Consequently, the regional government is currently testing Copenhagen's approach to its enhanced autonomy; sufficient consultation with the kingdom is still needed.

Greenland's approach so far has been to participate in existing forums and organizations, such as the AC, and to explore the potential of other co-operative venues (Arctic Frontiers, Northern Forum), rather than supporting the creation of or spearheading new co-operative multilateral mechanisms (Holm Olsen and Shadian, 2016). Greenland still looks east to establish ties as a Nordic country, especially in relation with neighboring Iceland and fellow devolved region, the Faroe Islands. On another note, resource development continues to dominate debates in Greenland, as bilateral ties with China and the European Union seem to be the primary conduit to tackle this controversial file (Lackenbauer *et al.*, 2018).

For a time, Alaska and the three Canadian territories were members of the Northern Forum, which acted as a potential forum for co-operation and governance; since 2013, all four, except Alaska, have withdrawn from the organization. However, some continued transnational work at the sub-national level can be found in bilateral relations across common boundaries, especially between Alaska and Yukon. For example, coordination agreements were signed between the two regional governments, which led to the 2012 *Alaska-Yukon Intergovernmental Relations Accord* and the memorandum of agreement that facilitated subsequent co-operative endeavours between the two administrations on such issues as expertise exchange, training, and health care (Government of Yukon, 2012). Sub-regional leaders have met in conferences spearheaded by civil society and have expressed the desire for further co-operation, especially on infrastructure development (Higginbotham and Spence, 2018; Higginbotham and Spence, 2018a). Additionally, Yukon, the Northwest Territories, and Alaska have founded the Arctic Caucus as part of the Pacific NorthWest Economic Region, which is an association fostering dialogue and coordination on cross-border issues. However, these initiatives are still at a nascent stage, if we compare them to the Barents co-operation.

Regional actors with global aspirations

New sub-national actors have entered the governance stage more recently, displaying a mix of functional and "identity paradiplomacy." As described above, identity paradiplomacy is characterized by the promotion of distinct values, interests, and identities, and pursues nation-building objectives. The governments of Québec and Scotland represent two examples of Arctic identity paradiplomacy, developing Arctic policies in response to rising global interest and efforts by non-Arctic states to court Arctic diplomatic connections and raise their circumpolar visibility.

Québec

The province of Québec was a late joiner to Arctic governance. The region is not even mentioned in the Government of Québec's 2006 International Policy Statement, and the province only entered the Arctic scene with the introduction of its Plan Nord (PN) in 2011, designed to develop the northern Nunavik region of the province (Roussel and Payette, 2011). The provincial plan aimed to:

> Ensure that resource development in the North is conducted in a way that benefits the population living in the area and contributes to progress in Québec as a whole. It is concerned that development in the area should be supported by local and aboriginal communities and be socially acceptable. It also wants the process to be part of an exemplary sustainable development approach that preserves the natural riches of the land and resources in the North.
>
> (Government of Québec, 2015: 12)

Critics have questioned the predominance of an economic development rationale over environmental or cultural matters (see Asselin, 2011; Gombay, 2013 among others), but the PN has marked the entry of the provincial government into a region previously dominated by federal and Indigenous government interest and attention (Arteau, 2017).

Global warming was a primary driver for Québec's Arctic engagement. As in other parts of the Arctic, thawing permafrost and increased sea levels will threaten infrastructure and heighten levels of human activity in this region (Roussel and Payette, 201). In general, the environment is a shared area of competency between the national and provincial governments in Canada. Moreover, provinces are responsible for implementing international treaties agreed upon by the federal government in Ottawa. Although Canadian provinces were consulted before and during international climate change negotiations, the Government of Québec adopted a hardline stance towards what it saw as the federal government's unsatisfactory and unambitious climate change position from 2006–2009 (Chaloux *et al.*, 2015: 304–306). Consequently, the province sent independent delegations to attend key international summits and conferences tackling climate change. This intensified international presence and highlighted Québec's support for a coalition of governments willing to lead on climate change mitigation.

Québec began to concentrate its efforts and resources on the Arctic Circle Assemblies in 2013. The province has participated in every summit since that time, organizing eight breakout sessions by 2018 with the assistance of the Institut Nordique du Québec (INQ). The 2014 session unveiled the PN, with subsequent panels focused on sustainable development, biodiversity, and the ecological and human impacts of climate change. The Québec panels featured a diverse array of bureaucrats, scholars, and Indigenous leaders exchanging views, and provincial cabinet ministers took part in panels on energy development in 2014 and 2017.

Québec's Premier Philippe Couillard spoke in plenary sessions in 2014, 2015, and 2016 (the last two on the opening plenary). Analysis of his speeches illuminates the province's normative goals. Environmental issues occupied two-thirds of the Premier's opening plenary allocutions, describing initiatives brought forward by his government and calling on global stakeholders to take more deliberate action. Climate change and the need for concerted efforts to mitigate greenhouse gas emissions occupied half of his speech, essentially preparing the terrain for the upcoming 2015 United Nations Climate Change Conference held in Paris in November/December 2015. Couillard's normative pitch was clear: pricing carbon emissions was pivotal to mitigate emissions, and the development of a cap-and-trade system offered the ideal remedy. The province also affirmed its commitment to protect Arctic biodiversity, committing to extending the number of areas protected from resource development.[3]

The concept of sustainable development in itself can include considerations of environmental protection and economic prosperity. Both objectives were part of Québec's government strategies for the North, although Couillard tended to refer to the environmental rather than the economic dimensions of the PN (mentioning environmental protection three times more often than economic development in his 2015 and 2016 Arctic Circle interventions). When speaking should read of economic development, Couillard explained how infrastructure (roads, airports, ports) could enhance investments, create jobs, and stimulate socio-economic development of Northern communities, yet allusions to specific economic projects were few and fleeting (Couillard, 2015, 2016). Additionally, the Premier's speeches encouraged the inclusion of diverse stakeholders in Arctic governance, framing collaboration and co-operation between governments, scientists, businesses, civil society, and Northern communities as inherently beneficial.

Provincial policy documents also reflected the new interest in the Arctic. Québec's 2017 International Policy Statement referred specifically to the Arctic region for the first time, presenting Northern and Arctic affairs from a sustainable development perspective that emphasized the transformative potential of climate change. The statement linked Québec's action to its Northern priorities: the PN, the inclusion of non-state actors (businesses, Aboriginal communities), and necessary scientific, cultural, and economic exchanges with Northern partners, including Scandinavian countries (Government of Québec, 2017: 49–50).

Scotland

The devolved government of Scotland adopted a similar approach to enter the international Arctic dialogue. In contrast to the Province of Québec, Scottish territory does not extend north of the 60th parallel (a standard geographic definition of "the North"). Instead, it is a sub-national government with considerable competencies, given that only international relations, defence, and management of non-renewable energy are the sole prerogatives of the United Kingdom Parliament.

Scotland's involvement in the Arctic region can be traced to their nation-building project and to identity paradiplomacy. Like Québec, Scotland has also joined the Arctic regional governance scene late, only entering Arctic forums in the early 2010s, and on many fronts its approach to Arctic issues has mirrored the activism of Québec.

The rise of the pro-independence Scottish National Party (SNP) served as a catalyst for this reorientation. In 2011, SNP-Member of Parliament Angus Robertson, called on Scotland to take a more active role in the Arctic region: "Constitutional developments in Scotland and significant environmental changes offer a real opportunity and imperative to properly engage with our wider geographic region," he proclaimed. "Our neighbours to the north and east have already made a good start and work constructively together. We need to join them and play our part. [. . .] The time has come to rediscover our neighbourhood and the issues, interests, opportunities and challenges we share" (Robertson, 2011). His views evinced strong antagonistic language in relation to London, accusing the United Kingdom of having "opted out of a serious approach" towards the Arctic. Since that time, turmoil over the Brexit issue has emboldened the Scottish independence movement's resolve that a more robust international presence is needed.[4]

In this context, the Government of Scotland took the Arctic Circle Assembly road to frame themselves as a Northern nation. The Scottish regional government started to engage with the Arctic Circle Assembly in 2016, with a keynote address by First Minister Nicola Sturgeon. Like Québec, she focused on global environmental issues by calling for bold action to implement the 2015 Paris Agreement and reduce greenhouse gas emissions, and championed the United Nations sustainable development goals. Both priorities stem from a globalist perspective in which Scotland aspires to be seen as a "good global citizen, a country determined the do right thing" (Sturgeon, 2016) – an outlook that monopolized half of Sturgeon's 2016 Arctic Circle address.

The Government of Scotland organized sessions during the 2017 and 2018 Arctic Circle meetings dedicated to environmental issues such as sustainable development and the transition to a low-carbon economy, culture, architecture, and literature. Similar to Québec, Scotland hosted an Arctic Circle Forum in November 2017, but it featured a more diverse array of topics: Internet connectivity, tourism, aviation, energy, transport, cargo shipping, and climate change. These reflect Scotland's Arctic agenda as laid out in its subsequent international policy documents.

The Scottish government issued its Nordic Baltic Policy Statement in September 2017, which outlined the history of engagement and "similar topographies, a mixture of urban and rural communities, and many similar socio-economic traits" between Scotland and Nordic countries (Government of Scotland, 2017a: 2). Scotland positioned itself as promoting progressive values as well as innovation and technology, praising Nordic countries for their human rights record, their innovation in energy production, and their expertise in digital technology. The document noted shared values and objectives on norms and standards, such as the development of a low-carbon economy and the implementation of a robust welfare model. Overall, Scotland framed itself as an open, modern, progressive society, advocating for green initiatives to ensure sustainable development and "committed to exploring opportunities for knowledge exchange, the sharing of best practice and the promotion of innovative join solutions with countries in the region" (Government of Scotland, 2017b: 10). With the exception of references to the Arctic Circle Assembly, the bulk of the Nordic policy was predicated on bilateral relations and initiatives with Nordic countries, with a special emphasis on Scandinavian countries in general and Norway in particular, revealing a narrower European Arctic focus rather than a general circumpolar policy orientation.[5]

Scottish government involvement in other Arctic subregional multilateral organizations has not yet materialized. For example, the West Nordic countries and governments might be obvious partners, given that the Scottish government framed their participation in the Arctic as part of their Nordic Baltic strategy. Regional forums such as the West Nordic Council (Iceland, Greenland, Faroe Islands) typically focus on recommendations and have low programmatic potential and implementation capacity (Stemre, 2018), thus limiting their interest for a sub-national actor seeking to raise its profile and visibility. The North Atlantic Cooperation (NORA), with members including Greenland, Iceland, the Faroe Islands, and coastal Norway, might prove an attractive opportunity because it builds networks between corporate, civil society, academic, and governmental stakeholders and directly funds sustainable development projects (which fits Scotland's Nordic policy). The inclusion of Norway, a key partner in Scotland's Nordic Baltic policy, may also add to its appeal.

Conclusion

Sub-national governments have been active players in Arctic governance for decades, driven by common interests. The Barents Euro-Arctic Region (BEAR) co-operation is a great example of interest convergence: regional governments from Nordic countries strove to engage with their Russian counterparts, while Russian regions looked for external partners to coordinate actions and provide financial aid for Arctic projects. By contrast, the heterogeneous membership at the Northern Forum inhibits a similar convergence of interests, limiting the potential of that multilateral forum. The international action of sub-national governments is typically rooted in pragmatic co-operation, with specific and concrete deliverables and/or interests. When such potential gain is absent, their participation in Arctic governance is jeopardized, as shown by sub-national governments exiting the Northern Forum. Québec and Scotland are additional examples of this dynamic. Both inserted the Arctic in their broader international strategy and strategic thematic priorities, such as climate change, and see regional forums (particularly the Arctic Circle) as tools to project an image of responsible global citizenship.

In terms of governance roles, sub-national governments have not played a significant regulatory role in the Arctic region. Global and regional institutions are the domain of sovereign states, leaving little space to sub-national governments. On the other hand, regional governments in federal states (such as Québec or Alaska) have exclusive or shared powers in key jurisdictions. They can then develop and implement innovative practices and norms on their territory and share these practices with Arctic stakeholders. The province of Québec adopted this strategy by promoting the socio-economic and environmental norms in its Plan Nord to Arctic audiences.

The appraisal is different when the focus shifts to programmatic roles. These sub-national governments have proven themselves able to pool resources together and fund programs capable of implementing concrete actions. The BRC and (to a lesser extent) the Northern Forum have reported pragmatic initiatives on health, education, infrastructure, environmental, and economic issues.

Generative and catalyst roles are more widespread practices in Arctic paradiplomacy. Sub-national governments bring stakeholders together and are able to facilitate co-operative ties. They are also quite productive at funding initiatives to generate knowledge about the Arctic people and environment. Sub-national governments located in the Arctic region, as well as newcomers (Québec, Scotland), have invested significant resources to develop scientific expertise and articulate coherent generative processes.

Notes

1 The following fits the definition of government entity: "a regional government or government of a state, province, republic, krai, oblast, autonomous okrug, administration of a city of federal significance, or federal government in situations where no regional structure exists in a nation" (Northern Forum, 2013).

2 As of 2018, member regions were Kainuu, Lapland, Oulu, North Karelia, Finnmark, Nordland, Troms, Arkhangelsk, Karelia, Komi, Murmansk, Nenets, Norrbotten, and Västerbotten.

3 The Government of Québec set the objective to exclude 20% of its North from economic development by 2020 (from a current figure around 12%) (Government of Québec, 2018). The convergence of Arctic affairs and environmental issues is reflected in biologist Jean Lemire's appointment as the Government of Québec's envoy for Climate Change and Northern and Arctic Affairs in September 2017. Climate change mitigation and initiatives to reduce plastic pollution in oceans were at the center of Lemire's diplomatic action. This emphasis makes sense as Québecers are likely to perceive the environment as a top priority rather than sovereignty or the economy (Roussel and Payette, 2011: 950).
4 The "No" victory in the independence referendum held in September 2014 did not stop nationalist ambitions in Scotland, even though the Arctic was not specifically inserted in Scotland's 2015 International Framework. The 2016 Brexit referendum gave more credence to Scottish nationalists as a strong majority of Scottish supported the "remain" option.
5 Scotland released an updated international policy statement in December 2017 which, for the first time, referred explicitly to the Arctic region – albeit occupying a distinctly lower priority than relations with the European Union, international development, investment, and tourism (Government of Scotland, 2017a).

References

Anderson, George. 2008. *Federalism: An Introduction.* Oxford: Oxford University Press.

Arteau, Jean-François. 2017. "Établir une Véritable Politique Arctique pour le Québec." *American Review of Canadian Studies* 47, no. 2: 225–234.

Asselin, Hugo. 2011. "Plan Nord: Les Autochtones Laissés en Plan." *Recherches Amérindiennes au Québec* 41, no. 1: 37–46.

Auchet, Marc. 2011. "Greenland at the Crossroads – What Strategy for the Arctic?" *International Journal* 66, no. 4: 957–970.

Bailes, Alyson. 2013. "Understanding the Arctic Council: A 'Sub-Regional" Perspective.' *Journal of Military and Strategic Studies* 15, no. 2: 31–49.

BEAC. 2017. "Environmental Improvements at Barents Hot Spots: Factsheet." www.barentsinfo.fi/beac/docs/HotSpotsFactSheet2017EN.pdf.

BEAC. 2018a. "Regional Working Group on Transport and Logistics." www.barentscooperation.org/en/Working-Groups/Regional-Working-Groups/Transport-and-logistics.

BEAC. 2018b. "Barents Environmental Hot Spots." www.barentsinfo.fi/hotspots/.

BEAR. 1993. *Declaration: Co-operation in the Barents Euro-Arctic Region. Conference of Foreign Ministers, Kirkenes, January 11, 1993.* Kirkenes, Norway: BEAR. www.barentsinfo.fi/beac/docs/459_doc_KirkenesDeclaration.pdf.

BEAR. 2014. "The Barents Programme, 2014–2018." www.barentsinfo.fi/beac/docs/Barents_Programme_2014_2018_adopted_2_June_2013.pdf. Last accessed May 7, 2019.

BEAR. 2015. *Joint Barents Transport Plan, September 2015.* Kirkenes, Norway: BEAR. www.barentsinfo.fi/beac/docs/Joint_Barents_Transport_Plan_2015_Final_ENG.pdf.

BEAR. 2018. *Meeting Minutes: 26th Meeting of the Joint Working Group for Health and Related Social Issues, March 22–23, 2018.* Kirkenes, Norway: BEAR. www.barentsinfo.fi/beac/docs/JWGHS_Kirkenes_22-23_March_2018_Minutes.pdf

Chaloux, Annie, Stéphane Paquin, and Hugo Séguin. 2015. "Canada's Multiple Voices Diplomacy in Climate Change Negotiations: A Focus on Québec." *International Negotiation* 20, no. 2; 291–318.

Couillard, Philippe. 2015. *Philippe Couillard, Premier of Québec, Arctic Circle. Arctic Circle Secretariat, October 16, 2015.* https://vimeo.com/142617950.

Couillard, Philippe. 2016. *The Honourable Philippe Couillard, Premier of Québec, 2016 Arctic Circle Assembly, October 7, 2016.* https://vimeo.com/187343024

Dubreuil, Antoine. 2011. "The Arctic of the Regions." *International Journal* 66, no. 4: 923–938.

Gombay, Nicole. 2013. "Oubliez la Faune et la Flore et Vivez Par L'Argent: Le Plan Nord, L'Économie et le Développement Nordique." *Géographie, Économie, Société* 15, no. 4: 327–344.

Government of Québec. 2015a. "Notre Gouvernement et ses Partenaires ont Positionné le Québec Comme Leader Scientifique pour L'Arctique." *Press Release*, October 16–17. http://premier-ministre.gouv.qc.ca/actualites/communiques/mois-en.asp?mois=10&annee=2015.

Government of Québec. 2015b. "The Plan Nord Toward 2035–2015–2020 Action Plan." https://plannord.gouv.qc.ca/wp-content/uploads/2017/05/Synthese_PN_EN.pdf.

Government of Québec. 2017. "Québec on the World Stage: Involved, Engaged, Thriving." www.mrif.gouv.qc.ca/en/ministere/politique-internationale.

Government of Québec. 2018. "Protection of the Environment." *Société du Plan Nord.* https://plannord.gouv.qc.ca/en/priority-actions/protection-environment-biodiversity-conservation/.

Government of Scotland. 2017a. "Scotland's International Policy Statement." December 8. https://beta.gov.scot/publications/scotlands-international-policy-statement-9781788514026/pages/0/

Government of Scotland. 2017b. "All Points North: The Scottish Government's Nordic Baltic Policy Statement." September 29. https://beta.gov.scot/publications/points-north-scottish-governments-nordic-baltic-policy-statement/.

Government of Yukon. 2012. "Yukon and Alaska Sign Collaboration Agreement." *Press Release*, October 11. http://www.gov.yk.ca/news/12-191.html.

Hasanat, Waliul. 2012a. "International Co-operation in the Northern Forum: Emerging New Norms in International Law?" *Polar Record* 48: 372–386.

Hasanat, Waliul. 2012b. "Diverse Soft-Law Co-operation Forms in the Arctic: Do They Complement or Contradict Each Other?" *International Community Law Review* 14, no. 3: 273–299.

Hasanat, Waliul. 2010. "Co-operation in the Barents Euro-Arctic Region in the Light of International Law." *Yearbook of Polar Law* 2: 279–309.

Higginbotham, John and Jennifer Spence. 2018. *The North American Arctic – Energizing Regional Collaboration and Governance*. Waterloo, Ontario: Centre for International Governance Innovation.

Higginbotham, John and Jennifer Spence. 2018a. *Energizing Regional Collaboration and Governance – CIGI Summit of North American Arctic Leaders*. February 3, Ottawa: Centre for International Governance Innovation.

Holm Olsen, Inuuteq and Jennifer Shadian. 2016. "Greenland and the Arctic Council: Sub-national Regions in a Time of Arctic Westphalianisation." *Arctic Yearbook*, 229–250.

International Barents Secretariat. 2018. "Barents Saga." www.barentscooperation.org/newsletter/Newsletter-2-2012/Barents-Freeway.

Joenniemi, Pertti and Alexander Sergunin. 2014. "Paradiplomacy as a Capacity-Building Strategy: The Case of Russia's Northwestern Sub-national Actors." *Problems of Communism* 61, no. 6: 18–33.

Keating, Michael. 2000. *Paradiplomacy and Regional Networking. Forum of Federations: An International Federalism*. Ottawa, Ontario: Forum of Federations. www.forumfed.org/libdocs/ForRelCU01/924-FRCU0105-eu-keating.pdf.

Keating, Michael. 2013. "Regions and International Affairs: Motives, Opportunities and Strategies." In *Paradiplomacy in Action: The Foreign Relations of Sub-national Governments*, edited by Francisco Aldecoa and Michael Keating, 1–16. London: Routledge.

Lackenbauer, P. Whitney, Adam Lajeunesse, James Manicom and Frédéric Lasserre. 2018. *China's Arctic Ambitions and What They Mean for Canada*. Alberta: University of Calgary Press.

Lecours, André. 2002. "Paradiplomacy: Reflections on the Foreign Policy and International Relations of Regions." *International Negoti*ations 7: 91–114.

Leland, Sigve and Alf Håkon Hoel. 2008. "Learning by Doing: The Barents Co-operation and Development of Regional Collaboration in the North." In *The New Northern Dimension of the European Neighbourhood*, edited by Pami Aalto, Helge Blakkisrud, and Hanna Smith, 36–53. Brussels: Centre for European Policy Studies.

Massie, Justin and Marjolaine Lamontagne. 2018. "Par-delà les Champs de Compétence: L'Affirmation du Québec en Matière de Guerre et de Paix." *Canadian Journal of Political Science* 51, no. 3: 573–598.

McHugh, James. 2015. "Paradiplomacy, Protodiplomacy and the Foreign Policy Aspirations of Quebec and other Canadian Provinces." *Canadian Foreign Policy Journal* 21, no. 3: 238–256.

Myrjord, Anne. 2003. "Governance Beyond the Union: EU Boundaries in the Barents Euro-Arctic Region." *European Foreign Affairs Review* 8: 239–257.

Northern Forum. 1994. "Resolution 23: In Support of Sustainable Development. Rovaniemi, September 16, 1994." www.northernforum.org/en/the-northern-forum/documents/category/16-archive.

Northern Forum. 2013. "The Northern Forum: Rules of Procedure." October 18. Resolution 160. Moscow: The Northern Forum.

Northern Forum. 2018. "Northern Forum Activity Report from November 2015 to March 2018." www.northernforum.org/en/the-northern-forum/documents/category/7-annual-reports.

Norwegian Barents Secretariat. 2018. "The Barents Secretariat's Funding Programme." https://barents.no/en/barents-secretariat-s-funding-programme.

Paquin, Stéphane. 2004. "La Paradiplomatie Identitaire: Le Québec, la Catalogne et la Flandre en Relations Internationales." *Politique et Sociétés* 23, no. 2–3: 183–202.

Pogodaev, Mikhail. 2016. *The Northern Forum's (NF) 2016 Observer Review Report*. 31 May. Tromsø, Norway: Arctic Council Secretariat. https://oaarchive.arctic-council.org/handle/11374/1881.

Robertson, Angus. 2011. "High Time to Join our Friends in the North and Face the Arctic Challenge." *The Scotsman*, November 29. www.scotsman.com/news/opinion/angus-robertson-high-time-to-join-our-friends-in-the-north-and-face-the-arctic-challenge-1-1987989.

Roussel, Stéphane and Jean-François Payette. 2011. "The Other Sovereignties: Québec and the Arctic." *International Journal* 66, no. 4: 939–955.

Shevchuk, Igor. 2014. "Regional International Co-operation in the Arctic & Subarctic Zone: A View from Karelia." *Arctic Yearbook* 3: 1–4.

Speca, Anthony. 2012. "Nunavut, Greenland and the politics of resource revenues." *Policy Options*, May 1, 2012. http://policyoptions.irpp.org/magazines/budget-2012/nunavut-greenland-and-the-politics-of-resource-revenues/.

Stemre, Morten. 2018. "Opportunities for Scotland in the New North." *Journal of the North AtlanticandArctic*,January.https://jonaa.org/content/2018/1/22/scotland-and-the-changing-arctic-lr6dd

Stokke, Olav Schram and Geir Hønneland (editors). 2007. *International Co-operation and Arctic Governance: Regime Effectiveness and Northern Region Building*. London: Routledge.

Sturgeon, Nicola. 2016. "Nicola Sturgeon's Keynote at the 2016 Arctic Circle Assembly." https://vimeo.com/187190681.

Tsui, Emily. 2016. "Revisiting the Northern Forum: Lessons from Alaska's Involvement." *Arctic Yearbook* 5: 281–287.

Young, Oran. 2002. "Can the Arctic Council and the Northern Forum Find Common Ground?" *Polar Record* 38: 289–296.

Young, Oran. 2005. "Governing the Arctic: From Cold War Theater to Mosaic of Co-operation." *Global Governance* 11: 9–15.

Zimmerbauer, Kaj. 2013. "Unusual Regionalism in Northern Europe: The Barents Region in the Making." *Regional Studies* 47, no. 1: 89–103.

4 Civil society in Arctic governance

Indigenous peoples' organizations, observers, and Arctic residents

Civil society is "the socio-sphere located between the family, the state, and the market" (Anheier and Themudo, 2002: 193) that stands outside the state (Nossal *et al.*, 2015: 97), including such groups as charities, Indigenous peoples' organizations, non-governmental organizations, professional associations, religious organizations, unions, and service organizations (World Economic Forum, 2018). For our purposes, this chapter focuses on Indigenous peoples' organizations, Arctic Council (AC) observers such as science groups, and Arctic residents as key components of civil society in the Arctic region. Traditionally, international relations theory has explained that most power in the international system belongs to states. There is good reason to define Arctic governance as statist, given that the preeminent regional body (the AC) counts only states as voting members. The countries that have territory and sovereign rights to ocean areas and continental shelves in the region get to define who is in the Arctic "club" and largely set the conditions under which non-state actors and non-Arctic states get to contribute. Nevertheless, it is impossible to understand the development of regional politics since the 1970s without taking into account the active roles of Indigenous peoples (for example, Tennberg, 2000; Keskitalo, 2004; Shadian, 2014).

International relations theories tend to assess the power of civil society actors in the information-gathering realm, where these actors inform states of "interests, beliefs and expectations" (Apeldoorn *et al.*, 2003: 21). These actors generate knowledge that states can use to create foreign policy, if actors persuade state decision-makers that an action is in the national interest. Towards this end, civil society can play a catalyst role by bringing together actors to offer solutions to persistent problems. For example, the scientists who produced the climate assessments of the Intergovernmental Panel on Climate Change influenced international relations by collectively documenting an issue that begged state action and convinced states that a response is necessary, but did not dictate the response or negotiate the approach. In the context of Arctic governance, the influence and role of some groups extends beyond catalyst functions to involvement in policy development, including the involvement of Indigenous peoples' organizations in AC task forces that have produced internationally-binding agreements.

This chapter interrogates three key questions to help better understand the role and influence of civil society actors in circumpolar governance, paying particular attention to the AC. Which civil society groups are well represented in Arctic governance in a formal sense, and why? How do these groups seek to influence outcomes? Which potential civil society groupings or interests do not have an active role in the AC, and why?

Indigenous peoples' organizations

Indigenous peoples' organizations are well represented in Arctic governance, and the direct representation of Arctic Indigenous peoples in the AC has set an important diplomatic precedent (see English, 2013 for a full discussion). The reasons for this representation reflect the significant role that transnational Indigenous activism played in creating the idea of a circumpolar region in the first place – and acknowledges how Indigenous peoples are not just stakeholders in Arctic governance but rights-holders on the basis of their peoplehood. Their influence goes beyond the normal expectations of a civil society actor – the AC is the only international institution in which Indigenous peoples have a form of membership alongside states.

Internationally active Arctic Indigenous peoples' organizations engage in cross-border politics as peoples with rights to self-determination. seeking to overcome the legacies of colonialism. Colonial interventions by the Arctic states and other civil society actors (from fur traders to whalers to mining and energy companies) encroached upon Indigenous homelands for non-renewable resource extraction (Mitchell, 1996; Berger, 1985), religious conversion processes (Balzer, 1999), extension of law and justice (Grant, 2010), education, and medical care (Shkilnyk, 1985). The effects of Arctic colonialism, as well as the rapid social and economic disruption that accompanied it, continue to reverberate through Arctic societies and are reflected in dismal socio-economic and health indicators in Northern Indigenous communities (see Mitchell, 1996; Shkilnyk, 1985; Irlbacher-Fox, 2009; Vitebsky, 2005).

The organizations

In the Nordic Arctic, the state borders that grew up around Saami territories served to catalyze cross-border Indigenous connections beginning in the 1950s. Indigenous peoples' organizations conceptualized the Arctic as a region that transected state borders (Hagtvedt Vik and Semb, 2013). The Saami Council, founded in 1956, represents 100,000 Saami people from Finland, Russia, Norway, and Sweden. Similarly, Inuit ("the people"), the Inuktitut-speaking peoples of the Arctic, founded the Inuit Circumpolar Conference (ICC) in 1977 (now the Inuit Circumpolar Council) to develop and articulate a shared voice as a transnational Indigenous people spanning four nation-states (the Kingdom of Denmark-Greenland, Canada, the United States, and Russia). Propelled by growing interest in Arctic natural resources (oil, gas, mining) and by global environmental movements (such as the anti-whaling movement) on Arctic communities (Shadian, 2014), ICC constitutes as a highly successful effort at political organizing that represents 160,000 people. Based on this model, Canadian actors (including ICC representatives) actively assisted

Russian Indigenous peoples to establish a post-Soviet Indigenous organization in that country (Wilson, 2007). The Russian Association of Indigenous Peoples of the North (RAIPON), founded in 1990, represents 270,000 Indigenous peoples from 41 different Russian groups.[1]

In negotiations to create the AC, policy-makers from Alaska argued that the Aleuts and Athabaskans should have the same status as other Indigenous peoples in the new institution (English, 2013: 225). As a result, three new Indigenous peoples' organizations were created to ensure that all Arctic Indigenous peoples had a body capable of representing their interests and taking up a position as a permanent participant. The Arctic Athabaskan Council (AAC), founded in 2000, is an umbrella group for 17 different member organizations representing more than 45,000 Athabaskan people spread across 76 communities in Canada and Alaska.[2] The Aleutian International Association (AIA), founded in 1998, is much smaller, representing 13 Aleutian communities in Alaska and Russia with a combined population of 8,897.[3] The Gwich'in Council International (GCI), created in 1999, also represents about 9,000 constituents in Canada and Alaska.[4]

Becoming "permanent participants"

How did these organizations gain their formal positions in Arctic transnational governance, and in the AC more specifically? The Arctic Environmental Protection Strategy (AEPS) accredited three Indigenous peoples' organizations as observers: the ICC, the Saami Council, and the USSR Association of Small Peoples of the North (later RAIPON). The effects of environmental pollution on their Indigenous homelands and their rights therein justified their involvement. The founding AEPS document, notably, lists their names ahead of the other observers (Arctic Environmental Protection Strategy, 1991), acknowledging that their interest and stake in regional environmental health are stronger and more direct than the non-Arctic observer countries, such as Great Britain, Poland, and West Germany. Accordingly, this immediate post-Cold War Arctic region-building effort made it clear that Indigenous peoples had special rights and a unique perspective to offer to regional co-operation.

Indigenous peoples' organizations actively worked towards this outcome. Initially, state delegations held AEPS meetings that excluded observers (including Indigenous peoples' organizations) because the Arctic states had no legal obligation to include these actors. Understandably frustrated with this situation, the Indigenous observers organized a side meeting with state delegations during the 1993 AEPS meetings in Nuuk, Greenland, in which they "characterized their exclusion as unwarranted and contrary to the spirit of co-operation in the circumpolar world" (Fenge, 2013: 11–12). Consequently, the Arctic member states allowed Indigenous peoples' organizations to attend most AEPS meetings and events (12). The following year, Denmark began hosting the Indigenous Peoples' Secretariat at a cost of roughly $2.5 million US per year to "assist and provide secretariat support functions to the permanent participants" (Arctic Council, 1998b, 2000b). No other non-state actor or observer received similar institutional support.

Building upon this foundation, Indigenous peoples' organizations successfully negotiated inclusion in the AC with special status that distinguished them from mere observers. In the 1995–1996 negotiations that led to the creation of the AC, Canada, Denmark, Finland, Iceland, Norway, and Sweden supported having Indigenous peoples' organizations as full members of the Council. The United States and Russia were hostile to the notion, believing that this would encroach upon state supremacy in international organizations and set a dangerous precedent. According to then-Canadian Minister of Foreign Affairs Lloyd Axworthy, the Americans and Russians eventually conceded that the Indigenous peoples' organizations could be second-tier members without voting rights, as long as military security was explicitly excluded from the AC's mandate (Axworthy, 2012). Traditional military security concerns were not a major driver for the creation of an AC, which allowed the compromise (interview 4). During key negotiations in 1995, Canadian leaders assured Indigenous peoples' organizations that the government of Canada would walk away from the negotiations if the United States would not support an acceptable status for Indigenous peoples. After deliberation, the groups accepted second-tier status, even though it was not the position some had hoped for (interview 2).

According to the AC's rules of procedure, Indigenous peoples' organizations hold the status of "permanent participants" (PPs) in the forum, second-tier members compared to the Arctic states, but with considerably more weight and influence than observers. Although only Arctic states formally vote on Council decisions, which are based upon "a consensus of the eight Arctic states" (Article 7), the PPs can participate in all AC activities (Arctic Council, 1998: Article 4 and 5), add items to the agenda (Article 12 and 19), propose the size of delegations (Arctic 13), and initiate projects (Article 26). Accordingly, PPs exert a strong influence on outcomes, and no major AC activities or decisions proceed without their support, thereby giving them *de facto* power in decision-making.

Indigenous peoples' organizations secured influence in the AC for four main reasons. First, some state policy-makers perceived that Indigenous peoples enjoy special status in their Arctic homelands as a norm (interview 12). As a key group of Arctic residents, some policy-makers suggested a moral obligation existed to extend a privileged position to these groups, consistent with their distinct political status in some states. If Indigenous peoples' organizations held a lower status than non-Arctic state and other non-state observers, the entire AC would be run by people from outside the region; Arctic state diplomats who work for the AC are typically based in southern locations, a scenario that would undermine the legitimacy of a forum designed to represent the interests and voices of the Arctic as a region. Second, Indigenous peoples' organization leaders insisted that they be represented in a distinct fashion (interview 1), viewing participation in international politics as necessary to achieve group goals (interview 13). Third, Arctic states saw a strategic advantage in granting distinct status to PPs. State delegations would have to include Indigenous peoples' positions in state policies even where these deviated from what the state perceived to be national interests if Indigenous peoples did not have a special membership. Allowing Indigenous peoples to represent their own interests would give states manoeuvrability if there was disagreement with state positions. Fourth, some Canadian policy-makers thought a legal obligation existed requiring that the voices of Indigenous peoples in the AC be included, given the recognition of Aboriginal rights in the Canadian constitution (interview 3).

Policy-makers from the United States and Russia raised the opposite concern. The bureaucrats feared that if Indigenous peoples' organizations became full members, their leaders could leverage their international influence to shape their home countries' positions and then advance their own positions, which would essentially give them two votes in the AC (interview 1). Furthermore, giving membership to Indigenous peoples' organizations could create a precedent for other international forums (interview 14). As full members, these groups might use the AC to air grievances against the superpowers and disrupt state agendas to promote their causes. The ICC, for example, had been critical of the United States, especially of the 1971 *Alaska Native Claims Settlement Act* and 1972 *Marine Mammal Protection Act*, which led to the collapse of the seal fur industry (interview 15). Ultimately, policy-makers from the United States and Russia accepted a unique AC status for Indigenous peoples, as long as their representative groups constituted a category below that of member states.

The politics of influence

Indigenous peoples' organizations have sought to influence outcomes in the AC by using their insider status to publicly rebuke opponents or to ally with AC states behind the scenes to achieve desired outcomes. Four examples illustrate this influence. First, the AC released its landmark *Arctic Climate Impact Assessment* (ACIA) report in 2004, which revealed tensions over the AC's policy-making role. Introduced in Chapter 1, the ACIA was a major climate change assessment that brought together thousands of scientific studies and pushed forward alarming facts about climate change, such as rapid Arctic warming and sea ice decline, that attracted global attention (Arctic Council, 2004: 10). The United States sponsored the project during the dying days of the Clinton administration (Arctic Council, 2000a), and the assessment took four years to complete. At the AC ministerial meetings in Espoo, Finland, in November, 2001, participants debated whether the assessment should have policy recommendations (Arctic Council, 2001). By April 2003, the recommendations section had ballooned into a policy document of "at least 30 pages, containing three major chapters" (Arctic Council, 2003). Plans were hatched for a special AC meeting in August 2003 to "create a link between the scientific work of ACIA and the policy document" (Arctic Council, 2003).

By August 2003, United States officials vetoed the policy document. Republican President George W. Bush championed a different political agenda than his predecessor. His senior advisors wanted the ACIA report to provide only scientific evidence and offer no recommendations for action. Domestic electoral considerations explain this decision. The Bush administration was leery of the impact that recommendations for climate action might have on the November 2004 United States elections, given that the president was sympathetic to the oil lobby and generally hostile to climate action (interview 2). In March 2004, ICC representative

Sheila Watt Cloutier testified about the political motivations of the situation and the severe impacts of climate change on Arctic peoples before the Senate Committee on Commerce, which shamed the United States into reversing its stance (Watt-Cloutier, 2004 and interview 2). The fact that a policy document exists today (albeit one much shorter than originally envisioned and released after the presidential election) reflects the ICC's successful advocacy efforts. In this case, a PP used insider knowledge of the AC to publicly pressure the United States and achieve their desired outcome, even though it lacked a vote in the organization.

In a second case, the ICC played a crucial role in blocking the European Union (EU) from becoming an accredited observer in the AC over opposition to the EU's ban on importing seal products. Sealing is a way of life and an important source of food and income for Inuit. Despite Canada and other Arctic countries' insistence that their seal harvest is lawful, sustainable, humane, and strictly regulated, the EU adopted a general ban on the importation and sale of seal products in 2009 in response to popular concerns about the methods used to kill seals. Under this EU regulation, the EU market accepts seal products only if a "recognized body" attests that the products result from hunts conducted traditionally by Inuit and other Indigenous communities or from hunts conducted for the sole purpose of sustainable management of marine resources on a non-systematic, not-for-profit basis (European Parliament, 2009). Although the regulation appears to accommodate Indigenous rights, Canadian Inuit groups forcefully argued that the European Economic Community's 1983 ban on harp seal pup skins (which allowed Inuit-derived products while banning all others) proved how a general ban effectively destroys the market for all seal products (CARC, 1986). Subsequent legal processes, which pitted public morality against the human rights of Arctic Indigenous peoples, has widened the divide between the two sides. The dispute provoked strong Inuit statements against the EU's alleged disregard for the rights and livelihoods of Indigenous peoples in the Arctic, which forced the ICC to publicly challenge the myths and misinformation that underlay the European position and to mount legal challenges through Canada and Norway against what it considered to be an unfair law. When attempts to resolve the issue bilaterally failed, Inuit groups questioned the legality of the ban under European Union laws (Simon, 2011; Court of Justice of the European Union, 2015), and backed Canada and Norway as they challenged the EU seal ban through World Trade Organization (WTO) dispute mechanisms (Fitzgerald, 2011; Hossain, 2013; Perišin, 2013; Sellheim, 2015).

The seal dispute had a direct bearing on PPs and, by extension, Canadian support for the EU's application for accredited observer status at the Arctic Council. While the EU had been an *ad hoc* observer since 1998 (originally as the European Community) (Wouters *et al.*, 2016: 103), the European Commission formally applied to become a "permanent observer" in 2008 and identified the change in membership as an immediate priority in its policy released that year (European Commission, 2008). The prolonged Arctic Council deliberations on observer applications have been documented elsewhere, with recent scholarship confirming that Canada and Russia harboured deep reservations about the EU's application. In Canada's case, Inuit groups strongly urged the government to oppose it (NTI,

2013). "As long as [the] European Union doesn't have the required sensitivity to the needs of northerners," Canadian Foreign Affairs Minister Lawrence Cannon asserted in 2009, "I see no reason why they should be [. . .] a permanent observer on the Arctic Council" (CBC News, 2009). Accordingly, when the Arctic Member States and Permanent Participants considered 14 applications for observer status at the Kiruna Ministerial Meeting in May 2013, the Arctic Council "received the application of the EU for observer status affirmatively" but ultimately deferred a final decision on implementation until the Council ministers agreed by consensus that all of the concerns of Council members regarding the EU application were resolved. For Canada and various Indigenous groups, this meant finding specific ways to address the EU seal product import ban.

Since that time, the EU has repeatedly affirmed its intent to respect Indigenous interests and rights, seeking regular dialogue with PPs to improve mutual understanding. Prior to the final ruling of the WTO Appellate court in May 2014 (which upheld the previous ruling that the EU seal ban was "necessary to protect public morals"), the European Parliament passed a resolution acknowledging "the wish of the inhabitants and governments of the Arctic region with sovereign rights and responsibilities to continue to pursue sustainable economic development while at the same time protecting the traditional sources of the Indigenous peoples' livelihood and the very sensitive nature of the Arctic ecosystems." The EU Foreign Affairs Council conclusions on "Developing a European Union Policy towards the Arctic Region," adopted on 12 May 2014, called on the EU to explore "appropriate ways of ensuring that the representatives of Arctic Indigenous peoples are informed and consulted on EU policies that may affect them." It also urged Canada to "use the current positive momentum in EU-Canada relations to help resolve the remaining issue so as to allow for the full implementation of the Kiruna decision regarding the EU's observer status as soon as possible before the next EU/Canada summit." The remaining issue, not mentioned by name, was the seal ban (Lackenbauer and Lalonde, 2017).

The WTO ruling, coupled with the Arctic Council's requirement for consensus, encouraged both sides to broker an agreement to overcome the political imbroglio. For the EU, this meant ensuring that Canadian Indigenous peoples were treated the same as any other Indigenous communities seeking access for their seal products in markets within the EU. The "Joint Statement by Canada and the European Union on Access to the European Union of Seal Products from Indigenous Communities of Canada," released in August 2014, set out "the framework for cooperation to enable access to the European Union of seal products that result from hunts traditionally conducted by Canadian Indigenous communities and which contribute to their subsistence." Under this agreement, both sides reinforced "the importance of preserving the traditional way of life of indigenous communities" and pledged to "continue to engage in regular dialogue with representatives of indigenous communities, with a view to increasing mutual understanding" (European Union, 2014). In its decision on the Joint Statement, the European Commission noted that, in return, Canada had agreed to "lift its reservations concerning the EU's observer status in the Arctic Council" (European Commission, 2014). Progress has been made on implementing this agreement since 2014 (Fisheries and Oceans Canada, 2015), proving the success of the ICC's preferred policy option. Here, a PP used its insider status to pressure AC allies behind the scene to adopt a particular position, achieving a desirable outcome beyond their formal powers.

While the first two examples show how Indigenous peoples' organizations have influenced agenda and outcomes in the Arctic Council, a third case illustrates that Permanent Participant status in the Arctic Council brings stable participation, even in the face of domestic political turmoil and absence of "home state" national support. In November 2012, Russia's Ministry of Justice deregistered RAIPON for several reasons: concerns about its domestic influence (Weber, 2012); fears about foreign influence within the organization (Digges, 2012); links between group members and Russian separatist movements (Nilsen and Pettersen, 2012); and RAIPON's opposition to oil production on its self-governing lands (Wallace, 2013). Deregistration meant that, according to the Russian government, the group ceased to exist. Nevertheless, the group continued to attend AC meetings (George, 2012), and the Council released a statement in support of RAIPON (which, ironically, Russia's chief Arctic diplomat endorsed) (George, 2012). The organization was registered in April 2013 after a slight adjustment in its membership (interview 17). Here, PPs allied with states and asserted their unique status while publicly rebuking a member state that stood in their way.

Fourth, PPs also seek influence through on-the-record naming and shaming at AC meetings (Wilson Rowe, 2018). At the Kiruna Ministerial Meeting in 2013, Athabaskan Chief Michael Stickman reiterated criticism about the lack of immediate action on black carbon and the continuous establishment of task forces rather than a concrete action plan. He specifically asked Russian Prime Minister Sergey Lavrov to explain why this was the case, given that the Russian government directly opposed action to curb the release of this air pollutant (Arctic Athabaskan Council, 2013). Likewise, the Swedish Saami representative used the home-turf ministerial to present a negative picture of Indigenous rights in Sweden:

> Here in the Giron area, industrial development – in particular mining – poses a tremendous challenge to the local Saami reindeer herding people. In our delegation, we have local reindeer herders with us that can witness to the fact that corporate responsibility can be a two-edged sword. The Swedish government says that we expect the mining industry to behave responsibly when operating in the Saami areas. But at the same time, the mining company says we have to follow Swedish law. Here in the Giron area, corporate responsibility equals no responsibility.
>
> (Saami Council, 2013)

While the PPs enjoy the right to participate fully in the work of the AC, they face practical problems that limit their ability to play their intended roles, including their actions to hold state representatives to account. These problems are often described as a lack of capacity – a concept tentatively defined by the Arctic Council's 2002 "Capacity Building Strategy and Pilot Project" as "the organizational and technical abilities that enable an organization or community to mobilize and direct resources to achieve their objectives in accordance with their values." According to a 2013 scoping paper produced by three North American think-tanks, active participation is best understood as participation in a literal sense. This is only possible if the PPs have the human and financial capacity to propose, conduct, and participate in the work of the Council, and to prepare for, attend, and follow up on the broad range of other Council activities (WDGF *et al*., 2013). If one accepts this interpretation, the spirit of full consultation implies that PPs must play a role in the processes leading up to formal decision-making. This conclusion is consistent with the *Framework for Strengthening the Arctic Council*, adopted at the 2011 Senior Arctic Officials meeting in Nuuk, which stipulates that "decisions at all levels in the Arctic Council are the exclusive right and responsibility of the eight Arctic States with the involvement of the Permanent Participants" (quoted in Dodds, 2015: 378).

In practical terms, the Permanent Participants lack the organizational depth and technical expertise (and the funding to develop them) to participate on the same level as the Arctic states and some observers. These capacity challenges reflect the uncertain financial situation of these Indigenous organizations, a shortage of personnel fluent in the technical issues discussed at some Working Group and Task Force meetings, competing demands on a small number of personnel to participate in Council activities, and the difficulty of recruiting and retaining the right people to conduct research and represent the organizations at Council meetings.[5] The principle of "full consultation" creates an entitlement for PPs that Arctic Council member states are expected to respect and to proactively safeguard and facilitate (Molenaar, 2012: 177), making capacity issues highly relevant across the board. Jim Gamble, Executive Director of the Aleut International Association, notes that every AC Ministerial Declaration since Ottawa has mentioned the importance of states supporting Permanent Participant capacity. His statement testifies to the broadly recognized value that the PPs bring to the work of the AC, yet leaves the problem of how best to support small organizations grappling with an ever-expanding portfolio of unresolved Arctic governance issues (Gamble, 2015; see also Coote, 2016).

The Arctic states have committed to addressing the gap between the Permanent Participants' responsibilities and their capacity to fulfill them. Canada organized several meetings to discuss options during its chairmanship from 2013–15, confirming that human resource difficulties were closely connected to funding problems. Building on this momentum, the Permanent Participants announced in 2017 the creation of the Álgu Fund, seeking to secure a $30 million endowment to "provide stable, reliable, and predictable funding on an annual basis to participating organizations to facilitate their work in support of the Arctic Council" (Staalesen, 2017).[6]

In contrast, non-Indigenous Arctic residents or settlers represent a civil society grouping not actively represented in the AC. Former Yukon premier Tony Penikett (2017) asserts that, given the norm that the AC should include the voices of people of the North, non-Indigenous residents should have representation from regional governments or the Northern Forum in order to make the council more democratic.[7] Contrary to this criticism, avenues for non-Indigenous residents' representation include the biennial Conference of Parliamentarians of the Arctic Region, which brings together members of regional Arctic parliaments to discuss matters of mutual importance and provide inputs into the work of the AC. For example, the September 2018 meeting issued recommendations around four themes: "digitalization of the Arctic," climate change, corporate social responsibility for companies operating in the region, and the social well-being of people living in the Arctic (Conference of Parliamentarians of the Arctic Region, 2018a). Eleven members of parliament from around the circumpolar north form a standing committee that meets an average of four times annually in a northern locale to exchange information (Conference of Parliamentarians of the Arctic Region, 2018c).

Observers

Other civil society actors, such as human rights groups, environmental organizations, and science advocacy bodies, have less prominent formal positions in the Arctic Council than Indigenous peoples' organizations. As observers in the AC, non-governmental organizations (NGOs) and other associations represent the interests of their members, mainly consisting of professionals and concerned individuals. (Non-Arctic states also can be observers, as discussed in Chapter 1.)

The organizations

Six actors were identified as observers in the Arctic Environmental Protection Strategy (the precursor to the AC): the International Arctic Science Committee, Poland, the United Kingdom, West Germany, the United Nations Economic Commission for Europe, and the United Nations Environmental Programme (Arctic Environmental Protection Strategy, 1991). Today, 13 states, 13 international institutions, and 13 non-governmental organizations are AC observers (Arctic Council, 2018). As observers rather than members, these groups can attend AC meetings with the permission of the Arctic Council member states, make comments when invited to do so, and participate in AC projects with state consent.

Seven of the AC civil society observers represent scientists (the International Council for the Exploration of the Sea, the Advisory Committee on Protection of the Sea, the Arctic Institute of North America, the Circumpolar Conservation Union, the International Arctic Science Committee, the International Arctic Social Sciences Association, and the International Union for Circumpolar Health). Two of the observers are environmental advocacy groups (Oceana, and the World Wide Fund for Nature-Global Arctic Program). One observer is a human rights organization (the International Federation of Red Cross & Red Crescent Societies), and two groups advocate for the rights of Indigenous peoples (the Association of World Reindeer Herders, and the International Work Group for Indigenous Affairs). The remaining observer is an educational institution: the University of the Arctic.

Table 4.1 List of Arctic Council observers per types of actor.

States (13)	France (joined 2000), Germany (1998), Italy (2013), Japan (2013), The Netherlands (1998), China (2013), Poland (1998), India (2013), South Korea (2013), Singapore (2013), Spain (2006), Switzerland (2017), United Kingdom (1998)
Intergovernmental Organizations and International Institutions (13)	International Maritime Organization (2019), Nordic Council of Ministers (NCM) (2000), Nordic Environment Finance Corporation (NEFCO) (2004), North Atlantic Marine Mammal Commission (NAMMCO) (2000), International Union for the Conservation of Nature (IUCN) (2000), OSPAR Commission (2017), Standing Committee of the Parliamentarians of the Arctic Region (SCPAR) (1998), United Nations Economic Commission for Europe (UN-ECE) (1998), United Nations Development Programme (UNDP) (2002), United Nations Environment Programme (UNEP) (1998), West Nordic Council (WNC) (2017), World Meteorological Organization (WMO) (2017), Northern Forum (NF) (1998)
Non-Governmental Organizations (13)	International Council for the Exploration of the Sea (ICES) (2017), International Federation of Red Cross & Red Crescent Societies (IFRC) (2000), Advisory Committee on Protection of the Sea (ACOPS) (2000), Arctic Institute of North America (AINA) (2004), Association of World Reindeer Herders (AWRH) (2000), Circumpolar Conservation Union (CCU) (2000), International Arctic Science Committee (IASC) (1998), International Arctic Social Sciences Association (IASSA) (2000), International Union for Circumpolar Health (IUCH) (1998), International Work Group for Indigenous Affairs (IWGIA) (2000), Oceana (2017), University of the Arctic (UArctic) (2002), World Wide Fund for Nature-Global Arctic Program (WWF) (1998)

Becoming observers

What explains this presence of observers? The considerable representation of scientific organizations reflects how knowledge about Arctic environmental threats served as the major catalyst that led to the creation of the AC (Tennberg, 2000 and English, 2013). Much of the AC's ongoing collaborative work occurs in the six working groups (see Figure 4.1), where scientists – from state governments, NGOs/civil society and the academic realm – are highly visible. For example, the final report of the landmark *Arctic Climate Impact Assessment*, released in 2004, contained contributions from more than 300 authors, most of whom were scientists

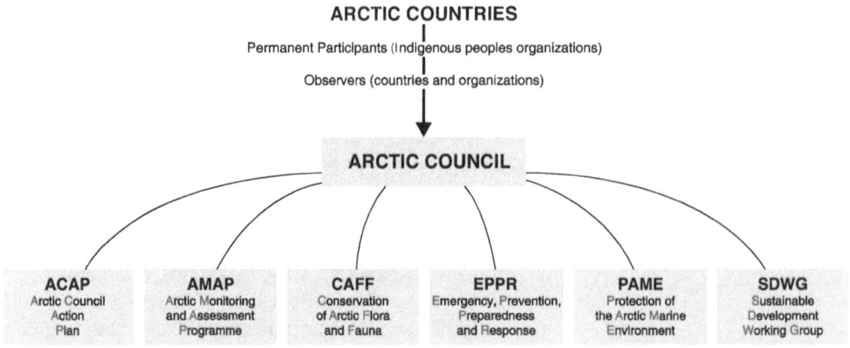

Figure 4.1 Arctic Council Working Groups

affiliated with universities, governments, or research institutes. Most of the AC's scientific observers contributed to the report, including some *ad hoc* observers. In short, even though foreign affairs departments administer the AC, scientists do most of the practical work at the project level.

Other civil society groups (such as human rights organizations, environmental organizations, and advocacy groups) are present in Arctic transnational governance as observers to the Arctic Council, where they attend meetings and participate in working groups. As Knecht (2016) revealed in his assessment of stakeholder group participation in the Arctic Council from 1998–2015, observer NGOs and intergovernmental organizations (IGOs) were the group with the lowest level of participation in AC settings, on average attending 25% (observer IGOs) and 27% (NGOs) of all Council events. Each organization sends relatively few representatives to meetings – usually one or two representatives, the fewest of any class of AC actors. Their participation also varied across settings. Observer NGOs attended 61% of all ministerial meetings and 54% of SAO meetings, only 19.5% of all WG meetings, and just 5.6% of Task Force meetings. These two last statistics are highly indicative, considering the working groups' and task forces' role in agenda-setting/policy option development as compared to the largely symbolic ministerial-level meetings. Observer NGOs and IGOs rarely make comments in AC plenary meetings (Chater, 2016, 2017) and rarely sponsor AC projects. Currently, the only project sponsor is the Nordic Environmental Finance Corporation (NEFCO), which is a Nordic governmental organization (see table 4.2).

Observer influence

Observers have limited avenues of influence. The formal constraints imposed on observer NGOs and IGOs clearly demarcate their status as tertiary actors in the AC, below states and Permanent Participants. "Observers may propose projects

Table 4.2 Project sponsorship in the Arctic Council for 2017.

Types of actors	*Actors*	*Projects Sponsored or Co-Sponsored*
States	Canada	31
	Denmark	16
	Finland	14
	Iceland	11
	Norway	36
	Russia	22
	Sweden	11
	United States	60
Permanent Participants	Aleut International Association	9
	Arctic Athabaskan Council	1
	Gwich'in Council International	4
	Inuit Circumpolar Council	5
	Russian Association of Indigenous Peoples of the North	1
	Saami Council	6
Observer	Nordic Environment Finance Corporation	2

Source: Arctic Council. 2017. *Amarok: The Arctic Council Tracking Tool Maxi-Report.* Fairbanks: 10th Arctic Council Ministerial Meeting.

through an Arctic State or a permanent participant but financial contributions from observers to any given project may not exceed the financing from Arctic States, unless otherwise decided by the [Senior Arctic Officials]," the AC explains. "While the primary role of observers is to observe the work of the Arctic Council, observers should continue to make relevant contributions through their engagement in the Arctic Council primarily at the level of Working Groups" (Arctic Council, 2018). Observers must assert their commitment and interest in the AC every four years in order to retain their status. They have to recognize the sovereignty of Arctic states as well as the "values, interests, culture and traditions of Arctic Indigenous peoples and other Arctic inhabitants" (Arctic Council, 2013). The early-established norm that the AC should include the voices of Indigenous peoples of the North precludes most observers, which generally consist of non-Indigenous residents and organizations with international interests, from attaining commensurate status and perceived legitimacy as the Permanent Participants.

Nevertheless, AC observers can influence Arctic states in informal ways. The World Wide Fund for Nature (WWF), an activist conservation organization, has sponsored events around AC meetings to raise awareness on issues of interest to NGOs. The WWF publishes reports on the state of Arctic conservation of polar bears and walruses, protection of Arctic marine areas, and the extent of regional ice coverage. In 2016, it pressured the AC by releasing a report card on how many of the AC's conservation goals have been implemented by states (WWF, 2016). AC meetings provide opportunities to get this research into the hands (and minds) of policy-makers. For example, during the Whitehorse AC meetings in October 2013, the WWF co-sponsored a day-long conference on corporate social responsibility in the Arctic. The AC holds its Sustainable Development Working Group meetings before Senior Arctic Officials meetings. The first day of AC meetings are closed-door affairs limited to top delegation officials, so events such as the WWF conference in Whitehorse provide an activity for policy-makers who have a free day between these meetings, thus affording NGOs with an important venue to share ideas and research. Given the constraints that these groups face, however, their focus on "open tent" venues such as the Arctic Circle Assembly (discussed in Chapter 1) remains understandable.

The level of influence by scientists and scientific organizations on the Council's work has varied over time and issue area (Koivurova, 2010; Nilsson, 2007; Wilson Rowe, 2018), as explored below. The policy influence of scientists in raising awareness about the impact of climate change on the Arctic was high, in that it contributed to a new framing of the Arctic as vulnerable to and a bellwether of climate change. Non-governmental researchers played a key role in the realization of a key AC report, *The Arctic Climate Impact Assessment* (ACIA), which represented a significant accomplishment. According to Koivurova (2010: 149–50), the assessment was instrumental in changing public perceptions of the Arctic region. Scientists documented the profound changes that had already happened or were about to happen. They framed the Arctic region as an early barometer of climate change for the rest of the world. Additionally, they portrayed the region as a space inhabited by numerous human communities, rather than as a barren wasteland, impacted by global phenomenon beyond their control.

Scientists who participated in the ACIA drew strong connections with global processes such as the Intergovernmental Panel on Climate Change (IPCC). They were also quite skillful into keeping the climate assessment distinct from the political sphere. Nilsson described the dissemination process masterfully (2007: 108): scientists had "a wish to have an intergovernmental platform to provide political legitimacy to the process in the region." The presence of all eight Arctic states and of Indigenous communities around the AC table certainly provided relevant stakeholder approval and diffusion for the assessment and recommendations. The AC itself received a great deal of public attention, as the release of the ACIA produced a peak in coverage (Chater and Landriault, 2016).

At the same time, researchers who spearheaded the ACIA were adamant about separating science and policy; the scientific report was distinct from the policy one, with the former being free from negotiations and state manoeuvring. In the case of the ACIA, clearly delineating science from policy generated a positive reception. Numerous assessments were initiated after 2004 to study the consequences of climate change on the Arctic environment and people; AC working groups reoriented their priorities to acknowledge this necessity (Koivurova, 2010: 150).

The 2007 Oil and Gas Assessment (OGA) is an interesting counter-example to the ACIA, representing a policy-relevant expert process that encountered a more

problematic reception. The report had been called for by the previous AC ministerial and focused on assessing the environmental, social, economic, and human health impacts of ongoing and projected oil and gas activities in the Arctic. While none of the SAOs contested the scientific findings of the assessment, some of them were concerned about the AMAP Chair's plans to release the overview report for public and policy circulation at the Arctic Frontiers conference in early 2008. These SAOs argued that the assessment summary included policy recommendations that went beyond scientific findings and would need to be cleared with national capitols in a political process, rather than a scientific one. The AMAP Chair replied that the summary document made only science-based recommendations, which had been subject to extensive national scientific review, and reminded those assembled that SAOs did not alter working group reports, but rather received them. Finally, the SAOs agreed to finish their review of the policy portion by mid-January 2007 and that the policy recommendations and executive summary would only be released if the concerns from governments were addressed prior to the planned launch at Arctic Frontiers (Arctic Council, 2007b: 11). At a later meeting, however, the SAOs were given the task the official launch of the OGA (Arctic Council, 2007a). In this highly socio-economically relevant field, some SAOs felt that the policy recommendations made by the working group were too prescriptive, or had somehow overstepped the mandate of normal procedure and division of labour between experts (observers and otherwise) and political actors in the AC.

Conclusion

The Arctic Council (AC) contains a mix of actors that are global in scope (such as certain science organizations and the World Wide Fund for Nature), as well as actors that are more regional in scope (such as Indigenous peoples' organizations). Civil society groups provide fresh understandings and perceptions on Arctic issues by promoting new ideas, sharing knowledge, and articulating issues. Although they do not wield the decision-making power of states, the groups exert influence by shaping how issues are framed. Indigenous peoples' organizations have shaped discussions about the policy-making role of the AC, their role in the institution, and which organizations have a right to be present. They have actively participated in the institution and performed state-like functions. Observers have pushed forward a science agenda in Arctic governance and made the Arctic region more international in scope, showing that actors far beyond the region have a significant interest.

Indigenous peoples' status as Permanent Participants (PPs) in the AC affords strong representation in Arctic governance. This unique position flows from their status and rights as self-determining peoples. Although such a role may seem self-evident, in no other international institution do Indigenous peoples have a similar status, having gained a formal position alongside states. Leveraging their influence in this consensus body where their status as the original Arctic peoples grants them considerable legitimacy, the PPs perform a regulatory function by participating in policy-making for the region. They are actively involved in co-leading and sponsoring projects, participating in working groups, and more. They have also wielded power in blocking applications for accredited observer status when the applicant's behaviour did not conform to their interests, as in the case of the EU ban on trade in seal products. As powerful as they are in the Arctic Council, in terms of their formal position, capacity and true consultation remains an ongoing challenge. Often, they turn to naming-and-shaming techniques and public pressure campaigns that are more in line with the usual powers of non-state actors in international decision-making. Given their important role in the AC, efforts to enhance PP capacity to actively and effectively participate in the forum are ongoing.

Science organizations represent the largest group of AC observers and play a significant role in AC projects. Other Arctic actors, such as environmental organizations and human rights groups, have much less influence in regional governance, based on their lack of membership in the AC. These groups provide states with useful information, and are thus important because they help states determine their interests. Climate change, for example, was brought to public light largely due to the efforts of scientists and partly due to institutions such as the AC.

Given this influx of observers to the AC, the Arctic member states have protected their privileged status, fearing that a growing number of observers with more clout could marginalize Arctic state and PP voices and render the AC unworkable (for example, interviews 3, 4 and 5; Kikkert, 2011). Yet, this has not occurred. Observers participate in the AC according to the rules. They contribute to projects but sponsor them only on a limited basis. They do not publicly rebuke member states or Indigenous peoples. Their presence in the Arctic region shows that civil society has an important role to play, but not a role that challenges the supremacy of the state.

Notes

1 Data on Permanent Participants gleaned from https://arctic-council.org/index.php/en/about-us/permanent-participants/.
2 See: https://arctic-council.org/index.php/en/about-us/permanent-participants/aac
3 See: https://arctic-council.org/index.php/en/about-us/permanent-participants/aia
4 See: https://arctic-council.org/index.php/en/about-us/permanent-participants/gci
5 The Saami Council and ICC are larger organizations, with broader mandates, comparatively strong funding, and larger staffs. GCI, AAC, and AIA were created specifically to serve on the Arctic Council and have smaller constituencies, a narrower mandate, and smaller staffs. The mandate and capacity of RAIPON is more difficult to gauge based upon available information. See Lackenbauer *et al.* (2014).
6 Indigenous peoples' organizations also send representatives to participate in the Arctic Circle Assembly and Arctic Science Ministerial Meeting, and each AC Permanent Participant selects three representatives to sit on the Arctic Economic Council (the same number that states get to select).
7 One could argue that national governments represent these Arctic residents, but these state actors also represent non-Arctic residents and national interests (as well as Indigenous peoples as state citizens). Concern that non-Indigenous Arctic residents would demand a special role in governance was a key reason that the United States initially opposed a special status for Indigenous peoples in the AC. One interviewee suggested that there was a

concern among United States policy-makers that Alaska would demand membership in the AC if groups such as the Inuit Circumpolar Council had membership (interview 14). By contrast English (2013) suggests that the primary American concern was that Indigenous peoples' organizations may seek similar status in other international forums.

References

Álgu Fund. 2017. "The Álgu Fund." www.algufund.org/.

Anheier, Helmut and Themudo, Nuno. 2002. "Organizational Forms of Global Civil Society: Implications of Going Global." In *Global Civil Society*, edited by Marlies Glasius, Mary Kaldor, and Helmut Anheier, 191–216. Oxford: Oxford University Press.

Apeldoorn, Bastiaan van, Henk Overbeek, and Magnus Ryner. 2003. "Theories of European Integration: A Critique." In *A Ruined Fortress?: Neoliberal Hegemony and Transformation in Europe*, edited by Alan W. Cafruny and Magnus Ryner, 17–46. Oxford: Rowan and Littlefield.

Arctic Athabaskan Council. 2013. *Remarks by Michael Strickman, Intenrational Chair, Arctic Athabaskan Council*. Tromsø, Norway: Arctic Council Secretariat.

Arctic Council. 1998. *Arctic Council Rules of Procedure*. Iqaluit, Nunavut: Arctic Council Ministerial Meeting.

Arctic Council. 2000a. *Notes from the Second Ministerial Meeting, Barrow, Alaska, U.S.A., October 12–13, 2000*. Tromsø, Norway: Arctic Council Secretariat.

Arctic Council. 2000b. *Report of Senior Arctic Officials to Arctic Council Ministers: Barrow, Alaska, United States of America, October 12, 2000*. Tromsø, Norway: Arctic Council Secretariat/Arctic Council Ministerial Meeting.

Arctic Council. 2001. *Arctic Council Senior Arctic Officials' Meeting, Espoo, Finland, November 6–7, 2001: Minutes*. Tromsø, Norway: Arctic Council Secretariat.

Arctic Council. 2003. *Arctic Council Meeting of Senior Arctic Officials, Reykjavik, Iceland, April 9–10, 2003: Minutes*. Tromsø, Norway: Arctic Council Secretariat.

Arctic Council. 2004. *Arctic Climate Impact Assessment: Impacts of a Warming Climate*. Cambridge, UK: Cambridge University Press.

Arctic Council. 2007a. *Arctic Council Meeting of Senior Arctic Officials, Tromsø, Norway, 12–13 April 2007*. Tromsø, Norway: Arctic Council Secretariat.

Arctic Council. 2007b. *Meeting of Senior Arctic Officials Final Report, 28–29 November 2007, Narvik, Norway*. Tromsø, Norway: Arctic Council.

Arctic Council. 2011. *Senior Arctic Officials Report to Ministers, Nuuk, Greenland, May 2011*. Nuuk, Greenland: Arctic Council Secretariat.

Arctic Council. 2013. *Arctic Council Observer Manual for Subsidiary Bodies*. Tromsø. Norway: Arctic Council Secretariat.

Arctic Council. 2017. *Amarok: The Arctic Council Tracking Tool Maxi-Report*. Fairbanks: 10th Arctic Council Ministerial Meeting.

Arctic Council. 2018. "Observers." *About Us*, January 17. https://arctic-council.org/index.php/en/about-us/arctic-council/observers.

Arctic Council Indigenous Peoples' Secretariat. 1998b. *Terms of Reference and Procedural Guidelines*. Espoo, Finland: IPS Board.

Arctic Environmental Protection Strategy. 1991. *Declaration of the Protection of the Arctic Environment*. Rovaniemi, Finland: Arctic Environmental Protection Strategy.

Axworthy, Lloyd. 2012. "Circumpolar Challenges Conference: Video of the September 27 Event." *Rideau Institute*, November 22. www.rideauinstitute.ca/2012/11/22/circumpolar-challenges-conference-video-of-the-september-27-event/.

Balzer, Marjorie M. 1999. *The Tenacity of Ethnicity: A Siberian Saga in Global Perspective*. Princeton, NJ: Princeton University Press.

Berger, Thomas R. 1985. *Village Journey: The Report of the Alaska Native Review Commission*. New York: Hill and Wang.

Canadian Arctic Resources Committee (CARC). 1986. "The Anti-Sealing Campaign." *Northern Perspectives* 14, no. 2. www.carc.org/pubs/v14no2/2.htm.

Canadian Broadcasting Corporation (CBC) News. 2009. "Canada Against EU Entry to Arctic Council Because of Seal Trade Ban." April 29. www.cbc.ca/news/canada/north/canada-against-eu-entry-to-arctic-council-because-of-seal-trade-ban-1.806188.

Chater, Andrew. 2016. "Explaining Non-State Actors in the Arctic Council." *Strategic Analysis* 40, no. 3: 173–184.

Chater, Andrew. 2017. "Trends Among the Arctic Council Observers, 2015–2017." *Polar Research and Policy Initiative*, August 4. http://polarconnection.org/arctic-council-observers-trends/.

Chater, Andrew and Mathieu Landriault. 2016. "Understanding Media Perceptions of the Arctic Council." *Arctic Yearbook* 5: 60–72.

Conference of Parliamentarians of the Arctic Region. 2018a. "13th Arctic Parliamentary Conference in Inari, Finland, September 16th to 19th 2018." *News and Activities*. www.arcticparl.org/announcements.aspx?id=6862.

Conference of Parliamentarians of the Arctic Region. 2018c. "SCPAR Meeting in Kiruna, Sweden 13th to 15th of May." *Standing Committee*. www.arcticparl.org/committee.aspx.

Coote, Michaela Louise. 2016. "Environmental Decision-Making in the Arctic Council: What Is the Role of Indigenous Peoples?" *Arctic Yearbook* 5: 30–47.

Court of Justice of the European Union. September 3, 2015. "The Court of Justice Confirms the Validity of The Regulation on Trade in Seal Products." *Press Release No. 93/15*. Kirchberg, Luxembourg City: Court of Justice of the European Union.

Digges, Charles. 2012. "Russia Strangles International Indigenous Peoples Organization as War on NGOs Continues." *Bellona*, November 15. https://bellona.org/news/russian-human-rights-issues/russian-ngo-law/2012-11-russia-strangles-international-indigenous-peoples-organization-as-war-on-ngos-continues.

Dodds, Klaus. 2015. "From Ilulissat to Kiruna: Managing the Arctic Council and the Contemporary Geopolitics of the Arctic." In *Handbook of the Politics of the Arctic*, edited by Leif Christian Jensen and Geir Hønneland, 375–387. Cheltenham, UK: Edward Elgar Publishing.

English, John. 2013. *Ice and Water: Politics, Peoples and the Arctic Council*. Toronto, Ontario: Allen Lane.

European Commission. 2008. "The European Union and the Arctic Region." *Communication from the Commission to the European Parliament and the Council, 20 November, COM(2008) 763final: 9*. Brussels, Belgium: European Commission

European Commission. 2014. *European Commission Decision on the Joint Statement by Canada and the European Union on Access to the European Union of Seal Products from Indigenous Communities of Canada, 18 August, C(2014) 5881*. Brussels, Belgium: European Commission.

European Parliament. 2009. "Regulation (EC) No 1007/2009 of the European Parliament and of the Council of 16 September 2009 on Trade in Seal Products." http://eur-lex.europa.eu/LexUriServ/LexUriServ.do?uri=OJ:L:2009:286:0036:01:EN:HTML.

European Union. 2014. *Joint Statement by Canada and the European Union on Access to the European Union of Seal Products from Indigenous Communities of Canada to the*

European Commission Decision, 18 August, C(2014) 5881. Brussels, Belgium: European Commission.
Fenge, Terry. 2013. "The Arctic Council: Past, Present, and Future Prospects with Canada in the Chair from 2013 to 2015." *Northern Review* no. 37: 7–35.
Fisheries and Oceans Canada. 2015. "Government of Canada Congratulates the Government of Nunavut on Important Step to Securing Access to European Markets for Seal Products." July 31. http://news.gc.ca/web/article-en.do?nid=1014449.
Fitzgerald, Peter. 2011. 'Morality' May Not Be Enough to Justify the EU Seal Products Ban: Animal Welfare Meets International Trade Law." *Journal of International Wildlife Law & Policy* 14, no. 2: 85–136.
Gamble, Jim. 2015. "The Arctic Council Permanent Participants: Capacity & Support: Past, Present & Future." *Arctic Yearbook* 4: 385–388.
George, Jane. 2012. "Arctic Council Officials Call for Reinstatement of Russian Indigenous Org." *Nunatsiaq News*, November 15. www.nunatsiaqonline.ca/stories/article/65674arctic_council_calls_for_russian_Indigenous_orgs_return/.
Grant, Shelagh. 2010. *Polar Imperative: A History of Arctic Sovereignty in North America*. Vancouver, British Columbia: D&M Publishers.
Hanne Hagtvedt, Vik and Anne Julie Semb. 2013. "Who Owns the Land? Norway, the Sami and the ILO Indigenous and Tribal Peoples Convention." *International Journal on Minority and Group Rights* 20, no. 4 (2013): 517–550.
Hossain, Kamrul. 2013. "The EU Ban on the Import of Seal Products and the WTO Regulations: Neglected Human Rights of the Arctic Indigenous Peoples?" *Polar Record* 49, no. 2: 154–166.
Interview 1 with Ray Arnaudo, former United States delegation member and Council chair, April 17, 2013.
Interview 2 with Terry Fenge, representative for AAC and former representative for ICC, February 21, 2013.
Interview 3 with former Canadian delegation member and Aboriginal Affairs official, winter 2013.
Interview 4 with former Canadian delegation member and current consultant, winter 2014.
Interview 5 with former Canadian delegation member and Yukon government employee, winter 2013.
Interview 12 with former United States delegation member and policy official, spring 2013.
Interview 13 with former United States delegation member and State Department official, spring 2013.
Interview 14 with former United States high-ranking delegation member and State Department official, spring 2013.
Interview 15 with Chester Reimer, Inuit Circumpolar Council consultant and delegation member, February 13, 2013.
Interview 17 with former Russian delegation member and government official, summer 2013.
Irlbacher-Fox, Stephanie. 2009. *Finding Dashaa: Self-Government, Social Suffering and Aboriginal Policy in Canada*. Vancouver, British Columbia: University of British Columbia Press.
Keskitalo, E. Carina H. 2004. *Negotiating the Arctic: The Construction of an International Region*. New York and London: Routledge.
Kikkert, Peter. 2011. "Rising Above the Rhetoric: Northern Voices and the Strengthening of Canada's Capacity to Maintain a Stable Circumpolar World." *Northern Review* no. 33: 29–45.

Knecht, Sebastian. 2016. "The Politics of Arctic International Co-operation: Introducing a Dataset on Stakeholder Participation in Arctic Council Meetings, 1998–2015." *Co-operation and Conflict* 52, no. 2: 203–223.

Koivurova, Timo. 2010. "Limits and Possibilities of the Arctic Council in a Rapidly Changing Scene of Arctic Governance." *Polar Record* 46, no. 2: 146–156.

Lackenbauer, P. Whitney, Adam Lajeunesse, Heather Exner-Pirot, and Jennifer Arthur-Lackenbauer. 2014. *Enhancing Permanent Participant Capacity at the Arctic Council: Options for Discussion*. Otterville: True North Consulting.

Lackenbauer, P. Whitney and Suzanne Lalonde. 2017. "Searching for Common Ground in Evolving Canadian and EU Arctic Strategies." In *The European Union and the Arctic*, edited by Nengye Liu, Elizabeth Kirk, and Tore Henriksen, 119–171. Leiden, Netherlands: Brill.

Mitchell, Marybell. 1996. *From Talking Chiefs to a Native Corporate Elite: The Birth of Class and Nationalism Among Canadian Inuit*. Montreal, Quebec: McGill-Queen's University Press.

Molenaar, Erik J. 2012. "Current and Prospective Roles of the Arctic Council System within the Context of the Law of the Sea." In *The Arctic Council: Its Place in the Future of Arctic Governance*, edited by Thomas Axworthy, Timo Koivurova, and Waliul Hasanat, 139–189. Toronto: Walter Duncan Gordon Foundation, Munk School of International Affairs and the Arctic Centre.

Nilsen, Thomas and Trude Pettersen. 2012. "Oslo: Moseev Case Raises Concern." *Barents Observer*, November 10. http://barentsobserver.com/en/society/oslo-moseev-case-raises-concern-10-11.

Nilsson, Annika. 2007. "A Changing Arctic Climate: Science and Policy in the Arctic Climate Impact Assessment." *Linköping Studies in Art and Science: Working Paper No. 386*.

Nossal, Kim Richard, Stéphane Roussel, and Stéphane Paquin. 2015. *The Politics of Canadian Foreign Policy*, 4th edition. Montreal: McGill-Queen's University Press.

Nunavut Tunngavik Inc. (NTI). 2013. "NTI Urges Canada to Deny the European Union's Application." April 25. www.tunngavik.com/blog/news/nti-urges-canada-to-deny-the-european-unions-application/.

Penikett, Tony. 2017. *Hunting the Northern Character*. Vancouver, British Columbia: UBC Press.

Perišin, Tamara. 2013. "Is the EU Seal Products Regulation a Sealed Deal? EU and WTO Challenges." *International and Comparative Law Quarterly* 62, no. 2: 373–405.

Saami Council. 2013. *Statement by Ms Áile Javo, President of the Saami Council, on the Occasion of the Eight Ministerial Meeting of the Arctic Council, Giron 15, May 2013*. Tromsø, Norway: Arctic Council Secretariat/Arctic Council Ministerial Meeting.

Sellheim, Nicolas. 2015. "The Goals of the EU Seal Products Trade Regulation: From Effectiveness to Consequence." *Polar Record* 51, no. 3: 274–289.

Shadian, Jessica M. 2014. *The Politics of Arctic Sovereignty: Oil, Ice and Inuit Governance*. Abingdon, UK: Routledge.

Shkilnyk, Anastasia. 1985. *A Poison Stronger than Love: The Destruction of an Ojibwa Community*. New Haven, CT: Yale University Press.

Simon, Mary. 2011. "Speech: The European Union, Canada, and the Arctic: Challenges of International Governance." September 23. www.itk.ca/media/speech/european-union-canada-and-arctic-challenges-international-governance.

Staalesen, Atle. 2017. "Establishment of a $30 Million Álgu Fund Marks New Beginning in Arctic Council, Indigenous Peoples Say." *Barents Observer*, May 12. https://thebarents

observer.com/en/life-and-public/2017/05/establishment-30-million-algu-fund-marks-new-beginning-arctic-council.

Tennberg, Monica. 2000. *Arctic Environmental Cooperation: A Study in Governmentality*. Burlington, VT: Ashgate Publishing.

Vitebsky, Piers. 2005. *Reindeer People: Living with Animals and Spirits in Siberia*. London: Harper Collins.

Wallace, Ron. 2013. *The Case for RAIPON: Implications for Canada and the Arctic Council*. Calgary, Alberta: Canadian Defence and Foreign Affairs Institute.

Walter & Duncan Gordon Foundation, The Oak Foundation and the McArthur Foundation (WDGF *et al.*). 2013. *Scoping Paper: Building Capacity Among the Permanent Participants*. Toronto, Ontario: Walter & Duncan Gordon Foundation, The Oak Foundation and the McArthur Foundation.

Watt-Cloutier, Sheila. 2004. "Written Testimony of Sheila Watt-Cloutier, Chair, Inuit Circumpolar Conference, Prepared for the U.S. Senate Committee on Commerce, Science, Transportation Hearing on Climate Change." *Inuit Circumpolar Conference Media and Reports*, March 3. www.inuitcircumpolar.com/index.php?ID=256&Lang=En.

Weber, Bob. 2012. "Russian Stuns Arctic Council by Suspending Aboriginal Group from Meetings." *Globe and Mail*, November 15. www.theglobeandmail.com/news/politics/russia-stuns-arctic-council-by-suspending-aboriginal-group-from-meetings/article5360638/?service=mobile.

Wilson Rowe, Elana. 2007. "Time, Idealization and International Development: Promoting Canadian Co-Management in Northern Russia." *Area* 39, no. 3: 323–330.

Wilson Rowe, Elana. 2018. *Arctic Governance: Power in Cross-border Relations*. Manchester, UK: Manchester University Press.

World Economic Forum. 2018. "Who and What Is 'Civil Society'." April 23. www.weforum.org/agenda/2018/04/what-is-civil-society/.

Wouters, Jan, Jed Odermatt, and Thomas Ramopoulos. 2016. "The EU and International Organizations." In *The Diplomatic System of the European Union: Evolution, Change and Challenges*, edited by Michael Smith, Stephan Keukeleire, and Sophie Vanhoonacker. London: Routledge.

WWF. 2016. "WWF Arctic Council Conservation Scorecard." http://wwf-ap.org/apps/acscorecard/.

5 Private sector actors and Arctic governance

This chapter explores the role of the private sector – businesses and business alliances – in contemporary Arctic transnational governance questions. Economic gain and the pursuit of livelihoods, both local and in networks of colonial empire, have long shaped regional politics. The Arctic has been and is a homeland for the Indigenous peoples and other local residents whose land-based and maritime economies provide for both self-sufficiency and trade. More southerly or distant markets have also shaped Arctic economies. For example, the pursuit of whales and other migratory marine mammals brought European whalers and hunters to the Arctic Archipelago of Spitsbergen (today known as Svalbard) from the 17th century onwards (Pedersen, 2006). Trading in furs and the quest to map the eastern edges of the Russian Empire pushed Russian actors towards the outer edges of the Eurasian Arctic in the 1730s and 1740s (Bonhomme, 2012; see Zaslow, 1971 for a study of similar dynamics in North America in the nineteenth century).

As modern nation-states integrated the Arctic, economic development served both economic and political purposes. Joseph Stalin prized industry and exploitation of the Arctic's natural resources as means to promote Soviet economic independence and prosperity in the "revolution from above" (Rowe, 2013; Bruno, 2016). Mining and metallurgy during the 1950s onward, followed by an oil and gas era in Siberia from the 1970s, became key strategic elements. Greenland offers a contemporary illustration of the interlinkage between economic pursuits and political goals; the country aspires to independence but, like many Arctic national/local economies, has a narrow economic base and is reliant on natural resources (Bertelsen *et al.*, 2015). Gaining full independence from Denmark will be highly dependent upon diversifying the economic base to secure sources of income beyond fisheries, hunting, tourism, raw materials, and land-based industries. The question of what kinds of economic development opportunities (large versus small scale; extractive versus renewable) should be pursued – along with identifying which actors should represent Greenlandic interests – are hotly debated in Greenlandic society (Nuttall, 2015; Jacobsen, 2018; Dyrendom Graugaard, 2018). Debates in the Canadian North and Alaska, where slow economic growth and lack of economic diversity contribute to boom and bust cycles, mirror this dynamic (for example, Glomsrød *et al.*, 2017).

Nationally specific conversations about economic development at the national and sub-national levels vary across the Arctic states. These interactions have distinct characteristics and histories that certainly shape the interests, information, and practices that states and business actors bring to the circumpolar governance table. This chapter sets aside this potential empirical focus, however, and instead directs attention towards the direct role of the private sector in Arctic transnational settings and concomitant debates, including longstanding Arctic Council (AC) discussions about sustainable development.

The various sites of public-private sector politics explored in this chapter – the AC, the Arctic Economic Council (AEC), and bilateral relations (Russia/Norway) – highlight how business actors' engagement in Arctic governance is part of and plays out in light of active international debates about how the Arctic should be used (and who it should benefit and how). The chapter also argues that other actor groups indicate a growing support for private sector engagement in Arctic regional governance. A brief review of Arctic economies and longstanding regional development debates indicates the nationally varied and complex backdrop against which private sector engagement occurs. A review of literature on the private sector in international relations suggests that the private sector's influence in global governance is primarily through self-governance, lobbying, provision of expertise, and a privileged position in the politics of anticipation (questions of future development). The chapter then returns to our Arctic case through taking three empirical vantage points: 1) the private sector (or mostly lack thereof) in AC work; 2) the formation of the AEC; and 3) the role of business in the broader Arctic "marketplace" of ideas about Arctic futures.

Contested development and "many Arctics"

The Arctic affects broader national economies across the circumpolar north in varying degrees. For example, the Russian North plays an outsized role in the Russian national economy. According to the latest estimates, oil and gas made up nearly 30% of the country's consolidated budget revenue and more than half of export revenue (Simola and Solanko, 2017: 4). 80% of Russia's natural gas and 20% of its oil are produced in a narrowly defined Arctic space; the inclusion of eastern and western Siberia significantly increased statistics (Simola and Solanko, 2017: 28). By contrast, Alaska contributed 0.3% to the United States' $17.3 trillion economy in 2014 (Speiser, 2015). While all Arctic economies are significant within their local contexts, it is easy to surmise that national-level representatives sent to the AC have different points of departure when it comes to priorities, degree of interest in, and specific concerns for Arctic economic activities. Therefore, key similarities and differences across Arctic national economies provide insight into the preferences and concerns brought to circumpolar, co-operative settings.

The contours of the main sources of revenue within domestic Arctic regions vary as well (see table 5.1). For example, Alaska and Arctic Russia share a strong economic reliance on natural resource extraction, while many of the other Arctic economies depend upon public administration and defence for employment.

Table 5.1 Top 3 value-added sources by industry, by % of gross regional product, in 2015.

Alaska	24.3% – mining and quarrying (19.3 of this is oil and gas)	21.1% – public administration and defence	10.5% – transportation and storage
Canada	19.1% – public administration and defence	17.2% – mining/ quarrying	10.8% – real estate activities
Faroe Islands	11.9% – real estate activities	11.4% – fishing	10.2% other services activities
Arctic Finland	16.6% – manufacturing (electrical/electronics, among others)	12.5% – health care and social work	12% – real estate activities
Greenland	17.9% – other service activities	10.5% – wholesale and retail trade	10.3% – construction 10.3% – public administration and defence
Iceland	16.7% – public administration and defence	12.6% – other service activities	8.4% – wholesale retail and trade
Arctic Norway (Nordland, Troms, Finnmark, and Svalbard)	18.3% – health care and social work	12.0% – public administration and defence	9% – education
Arctic Russia	51.7% – petroleum and other mining	7.6% – construction	6.4% real estate activities
Arctic Sweden	13.9% – other services activities	11.8% – health care and social work	11.7% – manufacturing

Source: Glomsrød *et al.* (2017).

Note: the "other services" category varies from country to country, but is a category usually consisting of smaller business initiatives (excluding construction, hotel, and food, which are covered by other categories, including the service sector [hairdressers, funeral services, and some tourism activities like excursions and specialized guiding]).

Accordingly, the various private sectors play a role in public and policy debates about the future of the region, namely in regard to what kinds of economic activity are possible and the forms of governance needed to support them. Therefore, industry actors are often present and active based on expected opportunities, as is discussed in greater detail below, as well as through established forms of activity. The main established economic industries are summarized in the table below, which demonstrates the diversity of economies of the North:

The circumpolar political discussions about and changing practices for including business actors (outlined below) are foregrounded by broader debates and uncertainties about how to balance conservation principles and economic development in the North. The contrary representations of the region as either a frozen nature preserve or a future resource frontier (and a range of representations in between these extreme visions of the Arctic) have long been debated (Kristoffersen and Langhelle, 2017; Steinberg *et al.*, 2015; Wilson Rowe, 2018). Associated with these broader representations of the Arctic are important questions of justice relating to whose interests development (or conservation) should serve (local communities? Arctic states? Global audiences?) (Jacobsen, 2018; Gerhardt *et al.*, 2019).

The ongoing balancing act between sustainable development and conservation policies and interests is evident at the AC. At the conclusion of Denmark's AC Chairmanship in 2011, the Greenlandic representative revealed some of the tensions. "The Arctic is not just about polar bears and ice," he stated. "What is most often missing from the discussion is the human aspect of the Arctic and the conditions in which we live" (Prime Minister of Greenland, 2011, translation from Danish). In discussing the Conservation of Arctic Flora and Fauna (CAFF)'s work (now focused on biodiversity challenges), several delegations at the Senior Arctic Officials (SAOs) meeting in Anchorage in 2015 asserted that CAFF needed to use both conservation and sustainable use as guiding principles. The SAO chair emphasized that use and conservation are not inconsistent, and suggested that the biodiversity conversation "may indicate a need to consider shifting emphasis and thus the Council's focus a little bit" (Arctic Council Secretariat, 2015b: 15).

The questions of what kind of economic development is appropriate for the Arctic – and the balancing act with environmental protection – is also debated amongst representatives of the Arctic states themselves. Discussing the oil and gas work of the AC during the American chair, Norway pointed out that the mature petroleum industry in Norway was an important "driver of regional development in northern Norway." The country's representative questioned whether the AC focused enough attention on the economic development value of oil and gas, as opposed to merely identifying the industry as a "source of environmental risk." Russia, Denmark, and Canada concurred, while other states retorted that "the substantial environmental consequences of oil and gas activity must be addressed" (Arctic Council Secretariat, 2016: 21–2). The session notes concluded with the observation:

> In general, delegates appear to want balance between environmental protection and sustainable development, but there is not yet broad agreement about what that means in this area. Russia and Norway, and any others who wish the AC to think more about economic and regional development, were asked to suggest concrete proposals on how the AC could focus more on the potential for O&G [oil and gas] as a driver of economic development, in order to help the AC consider whether activities could be expanded in that direction.
>
> (Arctic Council Secretariat, 2016: 21–2)

In short, business actors are engaging in a complex political situation, with the conservation versus economic development dichotomy and associated policy debates far from resolved at the international political level, and affecting the room to manoeuvre. Accordingly, how do representatives of business and industry navigate regional governance against this changing political landscape? To address this question, we turn first to the broader literature on the private sector in global governance to harvest some analytical cues, before returning to our circumpolar case.

Private actors in global governance

The private sector has had a large impact on many different policy fields, from the broad, complex fields of climate change mitigation and trade to more specific policy issues such as genetically modified organisms (GMOs), prevention of child labor, and regulation of financial markets. What businesses seek to achieve in global governance varies by field and policy issue. Orsini (2016) suggests a three-part typology based on a wide review of private sector engagement in global questions: 1) business actors may be against certain forms of regulation and seek to slow or redirect processes; 2) they might be passive and take a listening/learning mandate; or 3) they may be actively seeking to shape regulation or support international regulatory or other decision-making processes through lobbying, informational campaigns and provision of technical data and expertise, networking, or more direct forms of participation in global governance. This third point – on provision of expertise and networking/alliance building both internally amongst the private sector and in reaching out to other Arctic actor groups – tallies with much of the activity of business actors in regional Arctic governance.

The development of private sector "self-governance" represents an important manifestation of marshalling expertise and networking within industry sectors. Private sector actors can chose to work together outside the purview of states to develop voluntary codes and standards, self-financed certification regimes, and elective reporting within the private sector itself. For example, members of the Association of Arctic Expedition Cruise Operators (AECO) members developed a shared pledge to respect national and international laws, to apply additional measures to protect wildlife, to engage meaningfully with local communities, and to avoid introducing non-native species to the Arctic (AECO, 2018). Likewise, corporations are key players at the International Organization for Standardization (ISO) and work to establish common measurements, testing, and calibrations for particular industries. Standard development is "demand-driven" by business actors seeking homogenization, and requires specialized, in-depth knowledge about technical matters (Murphy and Yates, 2009). Standards are adopted at the ISO by national delegations including scholars, non-governmental organizations, government officials, and industry representatives. Such industry codes can evolve into broader generic standards, thereby influencing eventual outcomes in global governance through sector-specific self-regulation (Nasiritousi *et al*., 2016).

In the Arctic, the only industry to develop specific standards is the oil and gas sector. The British Petroleum (BP) oil spill in the Gulf of Mexico in 2010 cast doubt on the desirability of exploring and extracting oil in Arctic waters, especially in North America. Creating standards signalled to concerned governments that the industry writ large was adopting precautions to avoid a similar disaster in the Arctic region. Five of the six existing standards sets were developed from 2014 and onwards, with only one on offshore structures from 2010 (the year of the BP spill, but likely under development before the event as well, due to a sense of Arctic opportunity and associated risks). These Arctic-specific ISO standards on petroleum and natural gas focus on offshore structures/material requirements, risk management in cold workplaces, environmental monitoring and data collection, and ice management, thus dealing with safety issues related to platforms, equipment, and workers (ISO, 2019).

Expertise and experts from the private sector also can be found in global governance settings advocating for particular policy solutions. Business engagement on carbon capture and storage illustrates private sector expertise dedicated to promoting particular technological solutions. Nasiritousi (2017) notes that many oil and gas companies are present at United Nations Framework Convention on Climate Change (UNFCCC) meetings to follow debates and hold side events. A key area of policy that these private sector actors have sought to promote is carbon capture and storage, which is a technological solution allowing for the continued growth of the petroleum industry while reducing greenhouse gas emissions.

It is also essential to note that, most of the time, businesses intervene via larger coalitions or business associations, rather than as individual firms. Examples of this strategy include the AEC; the Global Industry Coalition, a voice for biotech companies; or how oil and gas companies have, within the climate policy field, been visible through the International Petroleum Industry Environmental Conservation Association or the World Business Council on Sustainable Development (Nasiritousi, 2017). The resultant politics that take place within formal organizations and informal coalitions are pluralistic, and internal conflict within coalitions and formal associations can weaken private sector influence (Meckling, 2011).

Finally, as this book argues, no one actor group shapes Arctic governance alone. Rather, these actor sets are engaged in collaborating with – or counteracting – other actor sets' claims and policy preferences. In other words, business actors are engaged in a wider field of elite politics (Seabrooke and Henriksen, 2017), often sharing professional backgrounds and life experiences with the heads of non-governmental organizations and state diplomats involved in global governance (in contrast to community organizers promoting attention to issues from a local level). For example, Equinor (formerly Statoil), which is 67% owned by the Norwegian government, has a "founding funder" leading role in the UN Global Compact Action Platform for Ocean Sustainability. The Norwegian government is the main convenor of this body (UN Global Compact, 2018), and the presence of both the Norwegian state and a predominantly publicly-owned Norwegian business demonstrates how states and the private sector can work in tandem at the international level. The incentives and practices of elite politics may also draw the private sector into global governance policy areas within which the company or sector has no specific interest to promote. Rather, major firms may join environmental governance processes when called upon to do so by key states or as part of their corporate social responsibility/public outreach policies, but lack the interest or policies necessary to become key stakeholders (Bled, 2009).

Mostly absent: business actors in the Arctic Council

The private sector has not been engaged to any substantive degree at the preeminent forum for transnational circumpolar issues – the Arctic Council – even though businesses are a target audience for much of the Council's work (most of which is about persuasion and best practices rather than enforceable regulatory initiatives). Economic development has factored into government statements on Arctic co-operation under various guises, from corporate social responsibility (CSR) under the Swedish chair to the formation of the Arctic Economic Council (AEC) under the Canadian chair (Wehrmann, 2017). Business actors are involved in other global governance forums, making their lack of presence at the AC particularly conspicuous.

A systematic review of the AC participation lists and Senior Arctic Official (SAO) final reports from 2009 and onwards reveal that there was no business presence or interventions in the formal meetings of the AC until 2018.[1] While invited guests from UNESCO and the IMO to the IASC and the University of the Arctic made presentations, business or industrial representatives never gave any statements at the ends of meetings under calls for "any other business" or within "observers and invited speakers" interventions at the conclusion of these meetings.[2] It was not until the third SAO meeting of the Finnish chair that AEC head Tero Vauraste became the first business actor included as a formal participant (Arctic Council Secretariat, 2018: 21).

This public absence from international politics does not, by extension, mean that Arctic states and business actors are excluded from policy conversations about positions and preferences on circumpolar multilateral issues. Political actors clearly consulted business actors prior to the establishment of the AEC. The political representative of the Task Force responsible for developing the "circumpolar business forum" (CBF), which eventually became the AEC, seemed apprised of industry preferences, noting that "business sees benefit in leaving it to them to work out how the details of the CBF would function" and also that "the Arctic Council needs to focus on how it will work with the CBF . . . what relation the circumpolar Business Forum should ultimately have to the AC, what would be its role and how members would be identified/selected/appointed" (Arctic Council Secretariat, 2013).

The SAO meetings are official diplomatic events. The countries involved assemble and register their delegations, with some limitations on delegation size, in remote Arctic locations. As discussed above, it is indeed unusual to see companies included in the formal delegations of any country (an exception being the inclusion of the Saudi Arabian oil company in the Saudi Arabian delegation to the UNFCCC), and therefore unsurprising that we do not see a lot of formal participation from businesses at the AC.

For these reasons, a focus solely on the diplomatic meetings would give a skewed view of the extent of business participation in AC work. Consequently, this chapter takes a closer look at the AC "Task Forces" as a kind of "most likely" case for business involvement. Would there be more participation from the private sector, given that the task forces are set up to secure rapid progress on new issues for the AC and do not have a formal diplomatic status? For several of the task forces, there is no participation data publicly available (the Task Force on Scientific Cooperation, which resulted in a binding scientific cooperation agreement in 2017; the Task Force on Arctic Marine Cooperation; the Task Force on Search and Rescue, which resulted in a binding search and rescue agreement in 2011; and the Task Force on Short-Lived Climate Forcers). The Task Force on Black Carbon and

Methane did assemble participation lists, and these lists indicated that no private sector actors were involved.

The two exceptions to business actors' non-participation/non-invitation in Arctic Council work are the Task Force on Arctic Marine Oil Pollution Prevention (TFOPP) and the Task Force on Telecommunications Infrastructure in the Arctic (TFTIA).[3] The TFOPP's minutes and records give the most insight into business actors' contributions to the process, and will be treated in-depth.

TFOPP was convened to continue the work of the binding agreement, signed in 2013, on oil-spill preparedness and response. The AC had worked on petroleum-related issues since 2002, had upheld a set of Arctic Offshore Oil and Gas Guidelines, had put forward recommendations through the Arctic Marine Shipping Assessment (Arctic Council, 2009), and had prepared numerous other soft-law instruments that emphasized best environmental practices in the industry. Companies involved in Arctic hydrocarbon extraction, such as Gazprom, Rosneft, Statoil, and Total, also developed oil-spill prevention standards for the Barents Sea that influenced the International Organization for Standardization (ISO)'s work on cold region petroleum and natural gas activities.

The main initiatives for the Task Force on Arctic Marine Oil Pollution Prevention were to identify how the Arctic Council could contribute to marine oil pollution prevention in the Arctic, recommend a concrete plan of action and, as appropriate, develop co-operative implementation (TFOPP, 2014). The group discussed a wide range of measures for petroleum activities, including support of industry standardization activities, exchange of experience/strengthening of regulatory cooperation, traffic monitoring and management, and reduction of risks associated with heavy fuel oil.

In terms of participation, business representatives attended at times as part of national delegations (the Russian delegation had participants from major state companies at each meeting and in one instance, Norway included a delegate employed at Statoil) and as "invited experts" at various junctures. Examples of business participants at various points in time include an Advisor to the President of Russian petroleum major Rosneft; a chief technologist on offshore rescue operations from Gazprom; the Association of Oil and Gas Producers (IOGP) with two OGP representatives from Statoil Arctic and Shell Government relations; a Senior Executive Vice President from Sovcomflot, the state-owned shipping company; and representatives of two Finnish companies working on engineering and management solutions for marine oil-spill pollution prevention and response (TFOPP, 2014). While this level of participation is substantial for business involvement compared to other AC settings, it is important to keep in mind that out of 40–65 total delegates attending each TFOPP meeting, only 5–10 participants came from the private sector.

The minutes of the meeting demonstrate that, although small in number, the business delegates provided expertise and input about the scope of the challenge and their preferred technological solutions when called upon. At the third meeting of the TFOPP, industry took on a role beyond that of listening and discussing briefs. The chair specifically invited representatives of the OGP (represented by employees of Shell and Statoil), and the International Association of Drilling Contractors (IADC, represented by an organizational employee), to present their experience in the field of prevention at the meeting. The content was as follows:

> The OGP representative gave a presentation on its different committees and task forces, which are involved in the activities connected with oil pollution prevention and regulatory work in the Arctic. The presenters proposed that the focus of work connected with oil pollution prevention should be on the appropriate performance standards, robust risk assessment methods, mitigation and implementation. The IADC representative explained the association's activities for Arctic drilling safety and environmental protection, including development of the IADC HSE Case Guidelines, its activities in the field of training and accreditation, and its participation in industry and government efforts. Both presenters remarked that industry has a great desire to share its knowledge with the AC.
>
> (TFOPP, 2014: 12–3)

At the final meeting, the Task Force invited industry members to speak again. The content of the expert presentations focused on the sharing of Arctic operating experiences and highlighted the measures (both technological solutions and best practices) the industry actors thought could best prevent oil pollution in the circumpolar region.

However, industry actors were not the only non-state group present. NGOs with environmental and Arctic regional expertise were also present, but any proposals made did not figure into the final, public record of the meetings. It seems industry experts and their expertise on "how to do" rather than "whether to do" Arctic oil and gas exploration were a better fit to the mandate of the Task Force, and thus figured into the TFOPP meetings and in the official record of these meetings. Indeed, these two actor groups seem to have been at odds. Wehrmann (2016), writing on NGO involvement in the TFOPP meetings based on interviews with the actors involved, notes that there was a fundamental disagreement between industry experts, the World Wide Fund for Nature (WWF), and the Circumpolar Conservation Union as to whether offshore oil and gas development in the Arctic was a responsible action at all. There was no open disagreement on this point in diplomatic settings, and no disagreement is noted in the official minutes of TFOPP and on the topic of whether to pursue oil and gas exploration in the Arctic at all. However, the divergent perspectives and agendas were discussed at the domestic level (Wehrmann, 2016), which again reminds us that non-state actors' likely use of the domestic "backstage" to influence positions taken by states in the transnational setting of the Arctic council.

An Arctic space for business? Situating the Arctic Economic Council

A broad review of business actor participation in AC work reveals that there has been little place for the private sector in the AC's formal diplomatic work thus far, except at the level of task forces in key issue areas (telecommunications and oil-spill pollution prevention). However, a great deal of formal diplomatic effort during the 2013–2015 Canadian chair of the AC focused on developing a transnational place for business, resulting in the Arctic Economic Council (AEC). What was the thinking behind this development, and what do the early days of an Arctic business alliance tell us about the role of the private sector in Arctic governance?

When Canada took the helm of the AC from Sweden in 2013, it set business and economic development as flagship emphases (see Chapter 1). All AC states and Permanent Participants supported the initiative (Quinn, 2016; Chater, 2017), which evolved through a Task Force to Facilitate a Circumpolar Business Forum. Early documents had envisioned a special place for state representatives in the new forum, but the resultant organization of, and for, business occupied an arm's-length position from the AC (Loukacheva, 2015). Distance between the work of the AC and a business forum had been an important requirement forwarded by the Inuit Circumpolar Council (ICC) (Arctic Council Secretariat, 2014). This ICC standpoint reflected the longstanding multilateral norm of holding business interests at an arm's length from AC activities (as discussed earlier).

Another point of contention surrounded what kind of businesses the emerging forum was meant to serve. Non-governmental organizations involved in Arctic affairs became concerned when the focus of the evolving business forum shifted to enrolling nationally significant corporations rather than supporting small and medium enterprises (Greenpeace, 2015; WWF, 2014). The World Wide Fund for Nature (WWF) had encouraged the "greater involvement of business in meaningful dialogue about sustainable Arctic development," but became concerned about transparency and accountability of a for-profit business organization operating independently from the AC (WWF, 2014). Some key Canadian voices, such as former foreign minister Lloyd Axworthy and former Circumpolar Ambassador Mary Simon, were apprehensive about the role given to private actors pursuing corporate concerns at a potential crossroads with the concerns of and directions given to the institution by participants in the AC (Axworthy and Simon, 2015).

A final debate within the minutes of the Canadian Chair's last SAO meeting in 2015 indicates why the creation of the AEC (as it came to be called) had been contentious (Arctic Council Secretariat, 2015a: 15–6). During the discussion, the ICC took the opportunity to chastise the assembled AC members: the "words sustainable and responsible have real meaning in their communities," and underlining that the AEC should not be involved in developing the AC's policies. Other delegates remained supportive of the initiative, with the Saami Council welcoming the AEC's work on Arctic stewardship, and the United States "noted that the Council has benefited from collaboration with industry for a long time and expressed hope that the AEC will support this by identifying industry experts to support the Council's work."

The AEC secretariat opened in Tromsø, Norway in 2015, marking the official start of work for the private sector in articulating a vision and focus areas for the new organization. The AEC describes itself as an independent body aiming to facilitate Arctic business-to-business relations and positioning itself as the preferred advisor to the AC on business issues. Membership is open to all businesses active in the Arctic, with special positions reserved for key industry representatives nominated by Arctic states and Permanent Participants of the AC. These representative companies range from the native corporations of Alaska, individual companies (Finland, Canada, Sweden, and the United States), shipping and ship owners' associations (Denmark, Finland, Norway, and Russia), complex holding companies based in the North, federations of industries and chambers of commerce (Iceland, Denmark, Norway, and Russia), and state-owned companies (such as Rosneft) (Chater, 2017; Arctic Economic Council, 2018a).

Businesses also can apply to be considered for Arctic or Permafrost Partners status. Akin to observers in the AC setting, these companies do not hold voting rights but can submit ideas and presentations. These partners can be businesses or Indigenous peoples' organizations coming from either Arctic or sub-Arctic regions. Permafrost partners from non-Arctic states are few, with only three non-Arctic companies hailing from Korea, Greece, and Germany (Arctic Economic Council, 2019a).

The AEC is further organized into several working groups, based on the business areas selected by the involved Arctic company representatives. Initially, the working groups were on infrastructure, maritime transportation, telecommunications, and traditional knowledge/stewardship (2016–2017). Five working groups were operating by 2019: maritime transportation, responsible resource development, connectivity, investments, and infrastructure and energy. Each working group welcomes AEC members, representatives from the business sector under scrutiny, leaders of local communities impacted, and external experts to contribute to the proceedings of the group. The chair of the working group provides leadership, with specific companies chairing to provide leadership and direction.[4] As is often the case in private sector governance, norm development follows the pace of corporate actors' involvement and interests in the process. The pace of each working group reflects the desire of the specific economic sector to develop norms and standards. For example, the connectivity working group was the first to publish its report in January 2017, entitled *Arctic Broadband – Recommendations for an Interconnected Arctic*. Since then, connectivity industry leaders also organized summits in 2017 and 2018, convening experts and stakeholders in order to build on the momentum of the report.

The AEC working groups carry out three key functions, which tell us something about what private sector actors hope to achieve in Arctic governance. First, they are mapping out baseline conditions and future opportunities within their own economic sectors. Information gathering initiatives include efforts to map Arctic energy resources, Internet broadband accessibility, shipping regulations, and hydrographic mapping. The connectivity working group, for example, issued a February 2017 report that documents the state of Internet connectivity in the Circumpolar North and offers recommendations on the best way forward (Arctic Economic Council, 2017b). The forum acts as a facilitator to increase co-operation between industry leaders.

AEC leadership and working groups emphasize a commitment to stringent regulatory regimes, indicating that their work is quite oriented towards developments in broader Arctic politics rather than solely internally directed towards relations and collaborations amongst the business actors themselves. In a commentary in the *Arctic Yearbook*, Tara Sweeney, then Chair of the AEC, and Tero Vauraste, then Vice-Chair, stated that the AEC aims to make the Arctic a "favourable place to do business," by working to remove trade barriers, forge circumpolar market connections, and ensure predictable and common-sense regulatory environments (Sweeney and Vauraste, 2016). The maritime working group of the AEC confirmed support for the regulatory regimes established by the AC. "Today's Arctic business environment is governed by stringent regulatory regimes . . . and so it should be – the Arctic is a pristine environment, where people have thrived for generations," it highlights. "The AEC will work to promote best regulatory practices and to the extent possible seek to align rules and regulations to ease the flow of business" (Arctic Economic Council, 2017a).

Second, business representatives are drawing up inventories of best practices for different industries, such as mining and energy extraction, that are active in the Arctic region in order to influence corporate behaviours and decisions. These AEC reports represent a private sector effort to demonstrate their ability to govern Arctic economic activity. The working group on investments and infrastructure, for instance, is developing a code of ethics for Arctic investments, which would act as a non-binding, voluntary standard-setting initiative promoting responsible investments in the region. As such, the code is imagined as "guidelines on what to take into account when investing in the Arctic, who to consult, and how to do it in a responsible manner" (Arctic Economic Council, 2018b). The working group stresses that efforts have to be invested in order to build a coalition of support around the code. This move, in turn, should help promote the code of ethics and make companies aware of the principles, which should ultimately boost the rate of adoption of these principles.

The AEC also serves as an advocacy body for corporate sectoral interests. Arctic actors perceive the AEC as "a synthesis of Arctic business perspectives for consideration by the Arctic Council" (Gwich'in Council International, 2018). All AEC working groups seek to promote recommendations to decision-makers, whether they are national government representatives or local communities' stakeholders and Indigenous rightsholders. The AC remains an ideal forum to inform these actors of corporate recommendations, despite the difficulty in determining how the two entities should interface (see the discussion of state preferences in Chapter 1).

The recommendations forwarded to the AC by the AEC working groups are often quite specific. For example, in the case of connectivity, recommendations focused on calling upon governments to provide tax incentives for corporations and developing further private-public partnerships in order to increase telecommunications infrastructure and opportunities for Northerners. The report stressed the need for close public-private collaboration when corporate ventures are not seen as profitable or realistic (Arctic Economic Council, 2017b). The maritime transportation sector proposes a harmonization of national policies to ensure "a model for regular safe navigation system," drawing on bilateral and multilateral arrangements observed in the Gulf of Finland, the Bering Strait, and the Saint Lawrence Seaway (Arctic Economic Council, 2017a).

Despite the AEC's active agenda to develop recommendations for decision-makers in the Arctic, the formal relationship of the AEC to the AC long remained to be worked out in policy and practice. A small step in clarifying the relationship was a recent Memorandum of Understanding signed between the AC and AEC, emphasizing the benefits of collaborating on issues of shared concern in the future (Arctic Economic Council, 2019c). As discussed above, these topics of shared concern are not solely about already extant Arctic economic sectors, but also about future opportunities. The management and development of Arctic futures is a further way that the private sector shapes Arctic governance, and the topic to which we now turn.

Private sector participation in the marketplace of Arctic governance ideas

Anticipated resource futures shape infrastructure investments, state policies, community life, and individual choices (such as career paths and education), even if the planned projects or avenues of economic development never come about (Dittmer *et al.*, 2011; Dodds, 2013; Wilson Rowe, 2017; Wilson *et al.*, 2017; Peyton, 2011; Glomsrød *et al.*, 2017). Intense anticipation of a region-specific petroleum (or blue economy, or shipping and so on) future is related to long traditions of Arctic political exceptionalism (e.g. that the Arctic will have its own unique future, distinct from global or national futures) (Dittmer *et al.*, 2011; Dodds, 2013). As this section explores through a case of Arctic bilateral relations around a petroleum opportunity, businesses – especially major industrial actors – play an important role in envisioning Arctic futures, thereby shaping Arctic governance through a more informal marketplace of ideas about the Arctic future.

The notion of the Arctic as a booming oil and gas province has become one of the strongest elite/policy representations of the region. Media and scholarly accounts frequently cite the United States Geological Survey's estimate that the Arctic contains 25% of the world's undiscovered petroleum reserves (Bird *et al.*, 2008). The positive visions of the resource-laden future are actively shared with local populations by economic actors, as part of corporate social responsibility and as part of the companies' regional self-profiling. The shale gas revolution in the United States, followed by the dramatic drop in oil prices since late 2014, led the sector to shelve many Arctic offshore oil and gas prospects, including the Shtokman field in the Barents Sea, as well as projects off the coasts of Alaska and Greenland. What are the consequences of economic anticipation – including the private sector's role – on Arctic governance?

The Shtokman case is illustrative, as it actively shaped relations between Norway and Russia and thus provides an instance of how business interests and transnational (in this case bilateral) political relations play out interactively. The Shtokman gas field, located in the Barents Sea some 600 kilometers north of the shores of the Kola Peninsula, is one of the world's largest natural gas fields. The field's development first garnered earnest attention in the early to mid-1990s, with letters of intent signed at various points with a variety of international oil companies, and shifting placement of Russia's state-owned enterprises as the lead company for the prospective project. In 2005, Russia signed co-operation agreements with Norway and France to develop the field, with Gazprom at the forefront. The agreement triggered an avalanche of bids from other international oil companies for field development (Claes and Moe, 2014).

What ensued has been described as a "beauty pageant" (Øverland, 2011), with companies seeking to present their competence in the best possible light. Interested parties highlighted corporate social responsibility and local relations, financing, technological capacity and expertise, and willingness to accept downward bargaining on contract terms. Eventually, Statoil and Total joined a consortium with Gazprom to create the Shtokman Development AG in 2008.

In anticipation of the Shtokman development to come, several changes ensued at multiple levels and on both sides of the Russian/Norwegian border. The prospect of exploration caused major shifts in the Norwegian state-owned industrial sector, including the merger of oil and gas majors Hydro and Statoil into one company. In the local communities, the development potential impacted residents; interviewees reported instances of residents of Murmansk acquiring untenable personal debt in anticipation of brighter times (Wilson Rowe, 2017). On the bilateral political level, the Shtokman prospect (which was seen as the first of many potential bilateral Russian-Norwegian petroleum collaborations) gave impetus to four-decade-long negotiations over a disputed maritime delimitation in the Barents Sea and Arctic Ocean. The delimitation negotiations concluded successfully in 2010, and Shtokman is considered a positive driver for the resolution of this important bilateral issue (Moe *et al.*, 2011).

The much-anticipated exploration never got off the ground, and the shareholder agreement expired in July 2012. The costly and challenging nature of the project made Shtokman a daunting prospect given oscillating global gas markets. Nevertheless, the imagined potential of the Shtokman gas project remained after market prospects for liquefied natural gas from the Barents Sea evaporated. The local Russian population, for example, was left with a negative assessment of foreign companies generally and a renewed appreciation of longstanding mining and other industries in the region, even if they still criticized their environmental track records or socioeconomic contributions. Regional authorities learned the hard lesson that acting as positively as possible to encourage businesses to come – and stay – was their primary contribution to challenges and opportunities in economic development in the region (for an extended discussion of these issues see Wilson Rowe, 2017).

In other words, the anticipatory practices in connection with the Shtokman field, which were driven forward by the preparatory activities and corporate social responsibility strategies of companies set to develop the field alongside government actors and media more broadly, have had lasting repercussions for how economic development and environmental risks are understood in the region. They have also framed understandings of what kinds of stakeholders matter for large-scale economic development. Such "ghost projects" and the private sector actors involved in them shape Arctic governance by forming the dispositions, policy trajectories, and stakeholder patterns that Arctic residents bring to new debates about circumpolar economic and social development.

Conclusion

What can be said about the role of the private sector in supranational Arctic governance? Certainly, the discursive and practical grounds are prepared for the involvement of business actors in terms of functional policy areas: interest in sustainable development and/or economic development has been a longstanding feature of cooperative circumpolar diplomacy. The establishment of the Arctic Economic Council is a concrete expression of a shared willingness, or at least a consensus, amongst Arctic Council members to have an organization in place for business actors in a regional, circumpolar body. However, the relatively long period of discussion and uncertainty about how the AC should relate to the AEC, and the absence of business actors as participants or invited delegates at senior arctic official-level meetings of the Arctic Council until 2018, suggests that the normative grounds for private sector engagement are still under development. In other words, the question of how these private actors should engage in circumpolar governance remains an open one.

The relative absence of business actors also reflects the conservation and environmental concerns that have long been at the heart of AC work, with advocates for these areas of emphasis conceptualizing business participation as unnecessary or unhelpful. Furthermore, the economies of the Arctic are highly varied and Arctic states differ in their representations of for whom, and for what, the region is governed. How they balance economic development and conservation is reflected in distinct legislation and regulations. In this light, particular industries were less likely considered an area of collective circumpolar concern, and instead as an object of national-level policymaking and discussion.

The recent involvement of company representatives or business alliance representatives in AEC task force work on telecommunications and oil-spill pollution prevention is therefore exceptional. The *ad hoc* nature of task forces offers flexibility in policy spaces with specific, industry-oriented mandates. The involvement and "invited expert" role of industry actors in these groups tallies with observations about businesses in global governance more generally. Despite popular misconceptions, businesses are rarely opposed to regulation, which is often a prerequisite for their activities. Instead, they participate as holders of expert knowledge and actively promote certain forms of regulatory or technological solutions over others.

This ties into AEC working group activities, which seek to establish baselines relevant to their industry (regulation, capital, human resources, needs, opportunity, existing actors) and then offer specific recommendations to and through the AC. Business actors also engage in self-governance, such as creating Arctic industry standards for cruise operators and promoting best practices/standards for Arctic oil and gas exploration in the International Organization for Standardization. Hence, by establishing best practices, private sector involvement can positively affect regulatory processes and practical applications, and demonstrates the private sector's willingness to regulate in support of particular economic opportunities.

How well the AEC will function and thrive as a transnational governance mechanism relates to the body's relationship to the AC and to individual states. When alliance politics internally succeed and the AEC working groups formulate recommendations, what will happen if they are not heard in, or fail to even reach, the AC? Will businesses still consider the AEC an alliance worth investing in, or will they reinvest their efforts in national and sub-national arenas? Whatever the case, the private sector plays a significant role in shaping the "marketplace of ideas" about Arctic futures. Industry voices have broad community-level and political consequences, underlining how representations of anticipated futures are products of power relations and ideological commitments. Thus, private sector actors are not only experts and self-regulators, but key players in Arctic future-making more generally.

Notes

1 It is not possible to trace the backgrounds of each delegate at each meeting, as the level of detail in the participant lists varies from chair to chair (for example, no affiliations are given for the Danish, Canadian, and Swedish chairmanships). However, the absence of business actors in the meeting sessions for which organizational affiliations are included in the participation list suggests that the absence of business actors is a regional Arctic norm. The evidence of a norm or standard of non-participation from business is further confirmed by a review of the twice-yearly final reports from SAO meetings. An exception is one business representative part of the US delegation under the 2007–2009 chair. However, he (Walter Parker) was a former member of the civil service, operating briefly as a consultant before (under the subsequent Danish chair) showing up affiliated with the Institute of the North in Alaska. In other words, he was not a representative of industrial interests.

2 Most of the invited speakers represent observer organizations (as discussed in Chapter 4), and it is therefore worth noting that there are no business alliances amongst AC observer organizations. The Association of Oil and Gas Producers, about which we will see more later, applied to be an observer but has had their application deferred indefinitely (personal communication, 2019).

3 Of particular note from the TFTIA is 1) the direct involvement of business participants (a slightly rotating cast from Russia, Greenland, Norway, Sweden, and the United States) at three different meetings (as opposed to the "invited expert" role mostly taken by the private sector in TFOPP); and 2) the group sought to "make contact" with the AEC for information sharing and possibly coordinating with the AEC working group on telecommunications infrastructure (TFTIA, 2016). The pragmatic outreach effort to increase efficiency on shared telecommunications issues represents an interesting moment of the policy field actors seeking to negotiate the relationship between the AC and the AEC, even while higher-level political direction on what this relationship should look like was absent (more on this in Chapter 1 and later in this chapter).

4 For example, the Norwegian company Equinor is leading the Energy working group, while the Canadian ATCO group – a utilities company – is chairing the Responsible Resource Development working group (Arctic Economic Council, 2019b).

References

Arctic Council Secretariat. 2009. "Arctic Marine Shipping Assessment 2009 Report." https://oaarchive.arctic-council.org/handle/11374/54?show=full. Last accessed September 6, 2019.

Arctic Council Secretariat. 2013. *Arctic Council SAO Meeting, Whitehorse, Canada, 22–23 October 2014, Final Draft Report*. Tromsø, Norway: Arctic Council Secretariat.

Arctic Council Secretariat. 2015a. *Arctic Council SAO Meeting, Whitehorse, Canada, 4–5 March 2015, Final Draft Report*. Tromsø, Norway: Arctic Council Secretariat.

Arctic Council Secretariat. 2015b. *Summary Report: SAO Plenary Meeting, Anchorage, Alaska, 21–22 October 2015*. Tromsø, Norway: Arctic Council Secretariat.

Arctic Council Secretariat. 2016. *Summary Report, SAO Plenary Meeting. Fairbanks, Alaska, 16–17 March 2016*. Tromsø, Norway: Arctic Council Secretariat.

Arctic Council Secretariat. 2018. *Report: SAO Plenary Meeting, Rovaniemi, Finland, 1–2 November 2018*. Tromsø, Norway: Arctic Council Secretariat.

Arctic Economic Council. February 17, 2017a. *Workplan: Arctic Economic Council Maritime Transportation Working Group*. Tromsø, Norway: Arctic Economic Council Secretariat.

Arctic Economic Council. 2017b. *Arctic Broadband: Recommendations for an Interconnected Arctic*. Tromsø, Norway: Arctic Economic Council Secretariat.

Arctic Economic Council. 2018a. "About Us." https://arcticeconomiccouncil.com/about-us/

Arctic Economic Council. 2018b. *AEC Investments and Infrastructure Working Group: Mandate*. Tromsø, Norway: Arctic Economic Council Secretariat.

Arctic Economic Council. 2019a. *Members*. Tromsø, Norway: Arctic Economic Council Secretariat.

Arctic Economic Council. 2019b. *Our Work*. Tromsø, Norway: Arctic Economic Council Secretariat.

Arctic Economic Council. 2019c. "The Arctic Council Signs Memorandum of Understanding with the Arctic Economic Council." https://arcticeconomiccouncil.com/the-arctic-council-signs-memorandum-of-understanding-with-arctic-economic-council/.

Association of Arctic Expedition Cruise Operators (AECO). 2018. Guidelines. www.aeco.no/guidelines/.

Axworthy, Lloyd and Mary Simon. 2015. "Is Canada Undermining the Arctic Council?" *Globe and Mail*, March 4.

Bertelsen, Rasmus Gjedssø Jens Christian Justinussen, and Coco Smits. 2015. "Energy as a Developmental Srategy: Creating Knowledge-Based Energy Sectors in Iceland, the Faroe Islands and Greenland." In *Handbook of the Politics of the Arctic*, edited by Leif Christian Jensen and Geir Hønneland, 3–25. Cheltenham, UK: Edward Elgar Publishing.

Bird, Kenneth J. *et al*. 2008. *Circum-Arctic Resource Appraisal: Estimates of Undiscovered Oil and Gas North of the Arctic Circle*. Washington, DC: United States Geological Survey.

Bled, Amandine J. 2009. "Business to the Rescue: Private Sector Actors and Global Environmental Regimes' Legitimacy." *International Environmental Agreements: Politics, Law and Economics* 9, no. 2: 153–171.

Bonhomme, Brian. 2012. *Russian Exploration, from Siberia to Space: A History*. Jefferson, NC: McFarland and Company.

Bruno, Andy. 2016. *The Nature of Soviet Power: An Arctic Environmental History*. Cambridge, UK: Cambridge University Press.

Chater, Andrew. 2017. "Understanding Canada's Role in the Evolution of the Arctic Council." In *One Arctic: The Arctic Council and Circumpolar Governance*, edited by P. Whitney Lackenbauer, Heather Nicol, and Wilfrid Greaves, 78–100. Ottawa, Ontario: Canadian Arctic Resources Committee.

Claes, Dag Harald and Arild Moe. 2014. "Arctic Petroleum Resources in a Regional and Global Perspective." In *Geopolitics and Security in the Arctic*, edited by Rolf Tamnes and Kristine Offerdal, 97–120. London: Routledge.

Dittmer, Jason, Sami Moisio, Alan Ingram, and Klaus Dodds. 2011. "Have you Heard the One About the Disappearing Ice? Recasting Arctic Geopolitics." *Political Geography* 30, no. 4: 202–214.

Dodds, Klaus. 2013. "Anticipating the Arctic and the Arctic Council: Pre-emption, Precaution and Preparedness." *Polar Record* 49, no. 2: 193–203.

Dyrendom Graugaard, Naja. 2018. "'Without Seals, there Are No Greenlanders': Colonial and Postcolonial Narratives of Sustainability and Inuit Seal Hunting." In *The Politics of Sustainability in the Arctic*, edited by Ulrik Pram Gad and Jeppe Strandsbjerg, 74–93. Abingdon, UK: Routledge.

Gerhardt, Hannes, Berit Kristoffersen, and Kirsti Stuvøy. 2019. *Saving the Arctic: Green Peace or Oil Riot?* New York: Routledge.

Glomsrød, Solveig, Gérard Duhaime, and Iulie Aslaksen (editors). 2017. *The Economy of the North 2015*. Oslo, Norway: Statistisk Sentralbyrå,

Glomsrød, Solveig, Ilmo Mäenpää, Lars Lindholt, Helen Mc Donald, Taoyuan Wei, and Scott Goldsmith. 2017. "Arctic Economies Within the Arctic Nations." In *The Economy of the North 2015*, edited by Solveig Glomsrød, Gérard Duhaime, and Iulie Aslaksen, 37–38. Oslo, Norway: Statistisk Sentralbyrå.

Greenpeace. 2015. "Looking Back: Canada's Arctic Council Chairmanship." www.greenpeace.org/sa/looking-bac-canadas-arctic-council-chairmanship.

Gwich'in Council International. 2018. "Arctic Economic Council." https://gwichincouncil.com/arctic-economic-council.

ISO-International Standards Organization. 2019. "International Organization for Standardization: When the World Agrees." www.iso.org.

Jacobsen, Marc. 2018. "Digging Sustainability: Scaling and Sectoring of Sovereignty in Greenland and Nunavut Mining Discourses." In *The Politics of Sustainability in the Arctic*, edited by Ulrik Pram Gad and Jeppe Strandsbjerg, 52–73. Abingdon, UK: Routledge.

Kristoffersen, Berit and Oluf Langhelle. 2017. "Sustainable Development as a Global-Arctic Matter: Imaginaries and Controversies." In *Governing Arctic Change*, edited by Kathrin Keil and Sebastian Knecht, 21–38 London: Palgrave Macmillan.

Loukacheva, Natalia. 2015. "The Arctic Economic Council: The Origins." *The Yearbook of Polar Law Online* 7, no. 1: 225–248.

Meckling, Jonas. 2011. *Carbon Coalitions: Business, Climate Politics, and the Rise of Emissions Trading*. Cambridge, MA: The MIT Press.

Moe, Arild, Daniel Fjærtoft, and Indra Øverland. 2011. "Space and Timing: Why Was the Barents Sea Delimitation Dispute Resolved in 2010?" *Polar Geography* 34, no. 3: 145–162.

Murphy, Craig and JoAnne Yates. 2009. *The International Organization for Standardization (ISO): Global Governance through Voluntary Consensus*. New York: Routledge.

Nasiritousi, Nagmeh. 2017. "Fossil Fuel Emitters and Climate Change: Unpacking the Governance Activities of Large Oil and Gas Companies." *Environmental Politics* 26, no. 4: 621–647.

Nasiritousi, Naghmeh, Mattias Hjerne, and Bjorn-Ola Linner. 2016. "The Role of Non-State Actors in Climate Change Governance: Understanding Agency Through Governance Profiles." *International Environmental Agreements: Politics, Law and Economics* 16, no. 1: 109–126.

Nuttall, Mark. 2015. "Subsurface Politics: Greenlandic Discourses on Extractive Industries." In *Handbook of the Politics of the Arctic*, edited by Leif Christian Jensen and Geir Hønneland, 105–127. Cheltenham, UK: Edward Elgar Publishing.

Orsini, Amandine J. 2016. "Do Non-State Perspectives Matter for Treaty Ratification and Implementation? The Case of the European Consultation on the Nagoya Protocol." *Environmental Policy and Governance* 26, no. 5: 377–393.
Øverland, Indra. 2011. "Russia's Arctic Energy Policy." *International Journal* 65, no. 4: 865–878.
Pedersen, Thorbjorn. 2006. "The Svalbard Continental Shelf Controversy: Legal Disputes and Political Rivalries." *Ocean Development & International Law* 37, no. 3–4: 339–358.
Personal communication with Elana Wilson Rowe. January 22, 2019.
Peyton, Jonathan. 2011. "Corporate Ecology: BC Hydro's Stikine-Iskut Project and the Unbuilt Environment." *Journal of Historical Geography* 37, no. 3: 358–369.
Prime Minister of Greenland. 2011. *Formanded for Naalakkersuisut Kuupik Kleist, on Behalf of Denmark, the Faroe Islands and Greenland, Arctic Council 7th Ministerial.* Tromsø, Norway: Arctic Council Secretariat.
Quinn, Eilis. 2016. "How will the Arctic Economic Council Shape Business Future of the North?" *Anchorage Daily News*, September 28. www.adn.com/arctic/article/how-will-arctic-economic-council-shape-business-future-of-the-North.
Rowe, Lars. 2013. *Pechenganikel: Soviet Industry, Russian Pollution and the Outside World.* [Doctoral Dissertation]. Oslo, Norway: University of Oslo.
Seabrooke, Leonard and Lasse Henriksen (editors). 2017. *Professional Networks and Transnational Governance*. Cambridge, UK: Cambridge University Press.
Simola, Heli and Laura Solanko. 2017. *Overview of the Russian Energy Sector*. Helsinki, Finland: Bank of Finland.
Speiser, Matthew. 2015. "How Much Each State Contributes to the US Economy." *Business Insider*. www.businessinsider.com/how-much-each-state-contributes-to-the-us-economy-2015-9?r=US&IR=T&IR=T
Steinberg, Philip E., Jeremy Tasch, and Hannes Gerhardt. 2015. *Contesting the Arctic: Politics and Imaginaries in the Circumpolar North.* London: I. B. Tauris.
Sweeney, Tara and Tero Vauraste. 2016. "Arctic Economic Council: Creating Parameters for Sustainable Economic Development in the Arctic." *The Arctic Yearbook* 5: 119–121.
TFOPP. 2014. *Document Portfolio.* Tromsø, Norway: Arctic Council Secretariat.
TFTIA. 2016. *Co-Chairs' Summary*. Tromsø, Norway: Arctic Council Secretariat.
UN Global Compact. 2018. "Business Action Platform for Sustainable Ocean Business." www.unglobalcompact.org/take-action/action-platforms/ocean.
Wehrmann, Dorothea. 2016. "Shaping Circumpolar Agendas: The Identification and Significance of 'Emerging Issues' Addressed in the Arctic Council." *Arctic Yearbook* 5: 90–103.
Wehrmann, Dorothea. 2017. "Non-State Actors in Arctic Council Governance." In *Governing Arctic Change: Global Perspectives*, edited by Kathrin Keil and Sebastian Knecht, 187–206. Basingstoke, UK: Palgrave MacMillan.
Wilson Rowe, Elana. 2017. "Promises, Promises: Murmansk and the Unbuilt Petroleum Environment." *Arctic Review on Law and Politics* 1: 1–9.
Wilson Rowe, Elana. 2018. *Arctic Governance: Power in Cross-Border Relations*. Manchester, UK: Manchester University Press.
Wilson, Emma, Anne Merrild Hansen, and Elana Wilson Rowe. 2017. "Imagining the Future: Local Perceptions of Arctic Extractive Projects that Didn't Happen." In *Arcticness: Power and Voice from the North*, edited by Ilan Kelman, 130–149. London: UCL Press.
WWF. 2014. "New Arctic Economic Council Could be a Missed Opportunity." www.panda.org/wwf_news/?218590/aec-missed-opportunity.
Zaslow, Morris. 1971. *The Opening of the Canadian North, 1870–1914*. Toronto, Ontario: McClelland and Stewart.

Conclusion

Governing complexity in the Arctic: past, present, future

The Arctic is undergoing environmental transformation forced by global climate change, bringing new challenges that threaten established business operations, traditional livelihoods, biodiversity, and extant infrastructure. The change also underpins new opportunities, such as increased shipping, and generates or heightens the engagement of new actor groups.

Politically, climate change has attracted the interest of non-Arctic states, from Europe to Asia, in the potential opportunities and risks associated with a changing Arctic. In turn, the Arctic coastal states have responded to this "outside" global interest by demonstrating their ability to govern the region peacefully amongst themselves, pursuant to the United Nations Convention on the Law of the Sea, the 2008 *Ilulissat Declaration*, and subsequent regional agreements on search and rescue, oil-spill response, and science co-operation. Arctic states have sought to demonstrate that they are the natural leaders on Arctic issues, engaging with a broader and increasingly large set of global actors as needed. The conclusion of the Central Arctic Ocean Fisheries (CAOF) Agreement is a case in point, with the Arctic coastal states (A5) and subsequent Arctic 5+5 formats representing groups of states with differing sovereign or international rights in the Arctic Ocean (Chapter 2).

Both new and old opportunities and challenges play out against a complex and dynamic political landscape fraught with historical legacies and sociopolitical change. The circumpolar assertion of Indigenous peoples' rights and sovereignty, particularly since the 1970s, has been instrumental in the growth of transnational politics, forcing states and other actors to recognize the latitudinal linkages of peoplehood that transect the North-South boundaries of states. Likewise, it is impossible to understand the breadth and form of regional politics without taking into account the rush of region building and positive, forward-thinking foreign policy profiling by most Arctic states after the end of the Cold War. Enduring social, political, and economic interests and relations in the region – as well as the activism of Indigenous peoples – have made Arctic co-operation possible, even against today's broader backdrop of strategic rivalry and suspended relations between Russia and the other Arctic coastal states on the global stage.

As the chapters in this book reveal, the number of institutions within and relating to the Arctic region has multiplied. The Arctic Council (AC) remains the preeminent governance body, having grown in mandate, policy-making role, and institutional capacity over the last two decades (Chapter 1). The Arctic Circle Assembly represents a new, informal gathering for a wider range of interested stakeholders (Chapter 1), while the Barents-Euro Arctic Council and Barents Regional Council provide structures for sub-national actors to influence international governance (Chapter 3).

The number of actors that seek a role in governance has increased concurrently. Sub-national actors, such as Québec (with jurisdiction over part of the Canadian Arctic) and Scotland (a non-Arctic jurisdiction), seek greater involvement in circumpolar affairs, either on the grounds of Northern development plans (Québec) or global issues impacting the Arctic such as climate change (Scotland) (Chapter 3). Non-state actors, including Indigenous peoples' organizations, scientific bodies, and environmental groups, seek to influence the AC (with varying degrees of success) and participate in forums such as the Arctic Circle (Chapter 4). The private sector seeks to influence governance through self-governance (generating new standards in industry groups and through the International Organization for Standardization) and building the Arctic Economic Council to represent business interests (Chapter 5).

A changing Arctic governance network – scholarly implications

The chapters in this book have examined different "kinds" of actor groups – both longstanding and new – in Arctic politics and taken stock of their positions and influence in a range of governance settings. Our findings build upon and expand the baseline of Arctic scholarship in a few key ways.

First, our results show that not all actors are equally well-positioned to shape outcomes in Arctic governance. Global governance is often conceived of as a flat playing field in which a wide range of actors can be involved in "networked" policy-making. Recent research in transnational policy networks, however, has highlighted that these are sites of co-operation as well as competition and hierarchy. Not all voices count equally – even if many different "kinds" of voices are present in any one governance field. Although this observation may appear somewhat banal when it comes to considering states' roles in international relations (see Wilson Rowe, 2018 and Wegge, 2011, for a consideration of hierarchy in an Arctic context), hierarchy between Arctic non-state actor groups had yet to be considered.

Second, our chapters show how the changing emphasis on different actor groups and the rise of new settings and institutions affects other forums and actor groups. Chapters 1 and 2 explore what can be called the "rise of the state" in Arctic politics. While states have always been key actors in Arctic regional politics, the *Ilulissat Declaration* in 2008 marked a new chapter in state-led activity in the Arctic region, including multiple binding treaties negotiated between the Arctic states. The temporarily intensified focus on the A5 as a political format after 2007 (see Chapter 2), and the subsequent return to the larger A8 format for most Arctic political work, demonstrate that the states are experimenting actively in Arctic governance with different forms of networking. The decline of the Northern Forum, previously an active site of post-Cold-War region-building, to an organization adrift from a pressing mandate and geographical focus, also illustrates change in status and function of actor groups and institutions over time. The Northern Forum also illustrates the peril of too much inclusiveness, which can lead to diverging interests and subsequent disengagement, especially amongst players with few resources to invest in international affairs. In short, our cases reveal how and why Arctic governance networks are neither stable over time nor flat in structure.

The most striking form of change in Arctic governance networks has played out against a regional/global dynamic in how the Arctic and its policy challenges are understood and managed. In the two decades following the end of the Cold War, Arctic governance tended to be a regional affair, with Arctic states and Arctic civil society actors elaborating norms, standards, and rules. A handful of organizations controlled regional and subregional governance, with the AC emerging as the primary forum to coordinate co-operation between Arctic actors. Arctic states and Permanent Participants occupy front stage at the AC, which has evolved to address a wider range of policy issues spurred by climate change and the growing economic potential of the region. What seemed like a purely regional scene hid global connections, with scientists and Indigenous organizations connecting regional issues with global phenomena (such as contaminants and climate change) and tapping into global networks to raise awareness and leverage broader international mechanisms to advance their agendas. Furthermore, global institutions such as the World Economic Forum, International Maritime Organization, and International Organization for Standardization adopted or adapted norms and rules to better manage particular areas of activity in the Arctic region, such as financial investments, shipping, and natural resource extraction.

The tightening links between the Arctic region and the global system, as well as the future potential for even more integration, led a growing number of non-Arctic states to lobby for observer status at the AC. At the same time, Arctic states and sub-national actors established new meeting places to widen avenues of regional discussion: the Arctic Economic Council, Arctic Frontiers, and the Arctic Circle Assembly. These spaces, designed to be highly inclusive with little enforcement or regulatory function, emphasized more open exchange and discussion than the "Arctic club" mentality that some critics attributed to Arctic-state-dominated AC and A5 gatherings. Our analysis reveals that the integration of new players such as non-Arctic states and stakeholders at these new forums was selective and limited (see Chapters 1 and 5), even though media coverage suggested otherwise, and the new philosophy of openness led to an over-representation of some stakeholders (such as scholars from the Arctic states at the Arctic Circle Assembly).

Third, our focus on influence has allowed us to distinguish between the aims of actor groups and the limits that different actors have encountered in Arctic networked governance. Our "bird's eye view" of actors, including both more active and less active actor sets, helps capture the relative positioning of various actors vis-à-vis one another. This serves as a supplement to the wealth of literature on Arctic regimes (for example, Young, 2005; Koivurova and Vanderzwaag, 2006; Byers, 2012; English, 2013; Koivurova, 2010). Extant literature has captured how different actor groups have been instrumental in shaping certain outcomes in key policy areas (for example, English, 2013; Fenge, 2013; Gamble, 2015; Coote, 2016; Knecht, 2016), but has not theorized about the absence or decline of certain actor groups. Understanding the absence of certain potential actors and/or lack of influence offers important "non-findings" that help us better understand how Arctic politics and governance networks actually function. Likewise, our effort to capture a broader array of relevant actor groups in one book provides a more robust basis to consider how these myriad actors interact in shaping Arctic outcomes, building upon previous genealogies of particular Arctic actors and their Arctic politics (for example, Shadian, 2014; Nilsson, 2007; Riddell-Dixon, 2017; Wehrmann, 2016).

The chapters also illustrate that different actor groups have varying agendas and, naturally, varying degrees of success in realizing their aims. A plethora of actors play generative (creating new knowledge) and/or catalyst (bringing stakeholders together) roles in governance processes. In terms of agenda-setting (generative) capacity, an interesting finding of this book is that all engaged actors, whether state or non-state, Arctic or non-Arctic, promote preferred understandings about core challenges for Arctic governance. Some generate new bodies of knowledge to document changes in the Arctic environment and people.

Yet, the omnipresence of generative and catalyst actors did not lead to an increase in regulatory activity. The multiplicity of voices and forums, as well as the intensification of the scientific effort on specific issues, does not always create the political will to collectively employ new solutions, for example on climate mitigation. Furthermore, the sequence of policymaking varies. For example, in the case of the CAOF Agreement (explored in Chapter 2), regulations were adopted before generative work: gathering knowledge about the sustainable management of eventual Arctic Ocean fisheries will flow from the regulatory action.

Policy questions going forward

The Arctic is undergoing a "state change," as political scientist Oran Young put it over a decade ago. Does the political architecture need to undergo equally rapid change? We address a few potential questions below.

Is Arctic governance too complex?

We would argue that it is not, even though navigating complex networks of Arctic governance is a resource-demanding activity. Complexity can certainly exacerbate implementation gaps, as states and other actors can shop for a preferred forum and adhere to the standards most advantageous to them, or can go to the governance site most amenable to their broader interests. Nevertheless, the "large tent" of Arctic governance has its strengths. A regionally inclusive approach has produced innovative governance practices, such as the creation and entrenchment of the unique powers of Permanent Participants in the AC, where Indigenous peoples have a role comparable to states. It can also ensure that co-operation remains robust when the international political climate cools, whether over Russian aggression in the Ukraine (and corresponding Western sanctions) or changing American administrations' stances on climate change.

Is Arctic governance too "slow," and can governance keep pace with the changing political and natural environment?

Here, there is reason to worry. On the one hand, the preceding chapters have shown that stakeholders, especially Arctic states, are able to put in place institutions and arrangements to discuss future challenges. The agreement on the Central Arctic Ocean Fisheries is a key illustration of this capacity. The Arctic Council has evolved in many ways to become a broader institution and engages in dialogue with a range of actors, such as the private sector through the newly established Arctic Economic Council. On the other hand, the Arctic Council has been a key actor shaping discussion and elevating knowledge of the impacts of climate change in the Arctic region.

Nevertheless, agreement-creation remains a relatively rare occurrence in its output. Topics that might suggest international Arctic collaboration, such as black carbon, have not produced treaties or even strong shared policy statements owing to diverging national interests. The consensus-based approach to Arctic governance practiced in the Arctic Council has created a sense of shared purpose amongst Arctic states, but inevitably pulls the Arctic Council to the least-ambitious baseline, given the mixed and divergent policy preferences of the states involved. The Arctic Council states may consider having formats or spin-offs that allow a vanguard of engaged Arctic states to demonstrate what a collective, if not truly circumpolar, response to climate issues can look like. States have been protective of their role in the institution. This may be more effective than continuing to seek a fully Arctic consensus.

Do we need to expand the universe of Arctic governance settings?

The growing number of issue areas is likely to push in the direction of new bodies and some innovation. This is particularly notable, as the preferred method of Arctic governance (evidenced by the post-Cold-War political track record reviewed in Chapter 1) is to avoid creation of an overall regime or institution for the region. The outcome by default has been a proliferation of niche institutions fulfilling particular roles beyond the Arctic Council (such as the Arctic 5, Arctic 5+5, Arctic Economic Council, or Arctic Circle Assembly). The need for institutional creativity around the increasingly open and seasonally ice-free Central Arctic Ocean illustrates how material change in the Arctic region will push innovation (Balton, 2018). The CAOF Agreement represents the beginning, not the culmination, of the CAO governance process to establish a sustainable resource management system.

Does Arctic politics need to be more inclusive of non-regional actors?

As the region becomes integrated more closely in broader international flows, global governance institutions and actors will play an increasingly salient role in the construction and enforcement of Arctic rules. At the same time, the expansion of the geographical mandate of the Northern Forum and its decline in importance and relevance for sub-national actors is a cautionary point. A global Arctic forum may not be used by the actors – Arctic states and peoples – who have the most at stake and the strongest legal rights in Arctic governance. The negotiations and adoption of an international agreement to manage fisheries in the Central Arctic Ocean (CAO, see Chapter 2) among Arctic and non-Arctic states represents an interesting possible precedent for managing the balance between regional and global actors. In this case, Arctic states led the discussion and were able to suggest the most fundamental principles at the beginning of the process (through the *Oslo Declaration*). Then they involved non-Arctic states in the discussion under the Arctic 5+5 formula. The *Oslo Declaration* formed the basis of the time-limited moratorium on unregulated commercial fisheries in the CAO concluded by the Arctic 5+5. Whether this two-tiered, sequential diplomatic approach, with Arctic state agenda-setting followed by diplomacy with non-Arctic relevant states, will be replicated in the future remains to be seen. Whatever the case, Arctic governance systems must acknowledge and incorporate the rights and interests of a growing array of global stakeholders in a dynamic region that is not apart from, but a part of, an increasingly interconnected and globalized world.

References

Balton, David. 2018. "Arctic Science. Address at Arctic Circle Forum, Seoul, South Korea." December 7. www.wilsoncenter.org/article/david-balton-establishing-international-marine-science-organization

Byers, Michael. 2012. *Circumpolar Challenges: An Ambitious Agenda for the Arctic Council*. Ottawa, Ontario: Rideau Institute.

Coote, Michaela Louise. 2016. "Environmental Decision-Making in the Arctic Council: What Is the Role of Indigenous Peoples?" *Arctic Yearbook* 5: 30–47.

English, John. 2013. *Ice and Water: Politics, Peoples and the Arctic Council*. Toronto, Ontario: Allen Lane.

Fenge, Terry. 2013. "The Arctic Council: Past, Present, and Future Prospects with Canada in the Chair from 2013 to 2015." *Northern Review* no. 37: 7–35.

Gamble, Jim. 2015. "The Arctic Council Permanent Participants: Capacity & Support: Past, Present & Future." *Arctic Yearbook* 4: 385–388.

Knecht, Sebastian. 2016. "The Politics of Arctic International Co-operation: Introducing a Dataset on Stakeholder Participation in Arctic Council Meetings, 1998–2015." *Co-operation and Conflict* 52, no. 2: 203–223.

Koivurova, Timo. 2010. "Limits and Possibilities of the Arctic Council in a Rapidly Changing Scene of Arctic Governance." *Polar Record* 46, no. 2: 146–156.

Koivurova, Timo and David L. Vanderzwaag. 2006. "The Arctic Council at 10 Years: Retrospect and Prospects." *UBC Law Review* 40, no. 1: 121–194.

Nilsson, Annika. 2007. "A Changing Arctic Climate: Science and Policy in the Arctic Climate Impact Assessment." *Linköping Studies in Art and Science: Working Paper No. 386.*

Riddell-Dixon, Elizabeth. 2017. *Breaking the Ice: Canada, Sovereignty and the Extended Continental Shelf*. Toronto, Ontario: Dundurn Press.

Shadian, Jessica M. 2014. *The Politics of Arctic Sovereignty: Oil, Ice and Inuit Governance*. Abingdon, UK: Routledge.

Wegge, Njord. 2011. "The Political Order in the Arctic: Power Structures, Regimes and Influence." *Polar Record* 47, no. 2: 165–176.

Wehrmann, Dorothea. 2016. "Shaping Circumpolar Agendas: The Identification and Significance of 'Emerging Issues' Addressed in the Arctic Council." *Arctic Yearbook* 5: 109–121.

Wilson Rowe, Elana. 2018. *Arctic Governance: Power in Cross-Border Relations*. Manchester, UK: Manchester University Press.

Young, Oran R. 2005. "Governing the Arctic: From Cold War Theater to Mosaic of Cooperation." *Global Governance* 11, no. 1: 9–15.

Appendices

Appendix A

The declaration on the establishment of the Arctic Council (Ottawa declaration) (1996)

The representatives of the Governments of Canada, Denmark, Finland, Iceland, Norway, the Russian Federation, Sweden and the United States of America (hereinafter referred as the Arctic States) meeting in Ottawa;

- **Affirming** our commitment to the well-being of the inhabitants of the Arctic, including special recognition of the special relationship and unique contributions to the arctic of indigenous people and their communities;
- **Affirming** our commitment to sustainable development in the Arctic region, including economic and social development, improved health conditions and cultural well-being;
- **Affirming** concurrently our commitment to the protection of the Arctic environment, including the health of Arctic ecosystems, maintenance of biodiversity in the Arctic region and conservation and sustainable use of natural resources;
- **Recognizing** the contributions of the Arctic Environmental Protection Strategy to these commitments;
- **Recognizing** the traditional knowledge of the indigenous people of the Arctic and their communities and taking note of its importance and that of Arctic science and research to the collective understanding of the circumpolar Arctic;
- **Desiring** further to provide a means for promoting cooperative activities to address Arctic issues requiring circumpolar cooperation, and to ensure full consultation with and the involvement of indigenous people and their communities and other inhabitants of the Arctic in such activities;
- **Recognizing** the valuable contribution and support of the Inuit Circumpolar Conference, Saami Council, and the Association of Indigenous Minorities of the Far North, Siberia and the Far East of the Russian Federation in the development of the Arctic Council;
- **Desiring** to provide for regular intergovernmental consideration of and consultation on Arctic issues.

Hereby declare:

1 The Arctic Council is established as a high level forum to:

a provide a means for promoting cooperation, coordination and interaction among the Arctic States, with the involvement of the Arctic indigenous communities and other Arctic inhabitants on common arctic issues,[1] in particular issues of sustainable development and environmental protection in the Arctic.
b oversee and coordinate the programs established under the AEPS on the Arctic Monitoring and Assessment Program (AMAP); conservation of Arctic Flora and Fauna (CAFF); Protection of the Arctic Marine Environment (PAME); and Emergency Prevention, Preparedness and Response (EPPR).
c adopt terms of reference for and oversee and coordinate a sustainable development program.
d disseminate information, encourage education and promote interest in Arctic-related issues.

2 Members of the Arctic Council are: Canada, Denmark, Finland, Iceland, Norway, the Russian Federation, Sweden and the United States of America (the Arctic States).

The Inuit Circumpolar Conference, the Saami Council and the Association of Indigenous Minorities in the Far North, Siberia, the Far East of the Russian Federation are Permanent Participants in the Arctic Council. Permanent participation is equally open to other Arctic organizations of indigenous peoples[2] with majority Arctic indigenous constituency, representing:

a a single indigenous people resident in more than one Arctic State; or
b more than one Arctic indigenous people resident in a single Arctic State.

The determination that such an organization has met this criterion is to be made by decision of the Council. The number of Permanent Participants should at any time be less than the number of members.

The category of Permanent Participation is created to provide for active participation and full consultation with the Arctic indigenous representatives within the Arctic Council.

3 Observer status in the Arctic Council is open to:

a non-Arctic states;
b inter-governmental and inter-parliamentary organizations, global and regional; and
c non-governmental organizations

that the Council determines can contribute to its work.

4 The Council should normally meet on a biennial basis, with meetings of senior officials taking place more frequently, to provide for liaison and coordination. Each Arctic State should designate a focal point on matters related to the Arctic Council.

5 Responsibility for hosting meetings of the Arctic Council, including provision of secretariat functions, should rotate sequentially among the Arctic States.

6 The Arctic Council, as its first order of business, should adopt rules of procedure for its meetings and those of its working groups.

7 Decisions of the Arctic Council are to be by consensus of the Members

8 The Indigenous Peoples' secretariat established under AEPS is to continue under the framework of the Arctic Council.

9 The Arctic Council should regularly review the priorities and financing of its programs and associated structures.

Therefore, we the undersigned representatives of our respective Governments, recognizing the Arctic Council's political significance and intending to promote its results, have signed this declaration.

Signed by the representatives of the Arctic States in Ottawa on the 19th of September, 1996.

- For the Government of Canada
- For the Government of Finland
- For the Government of Norway
- For the Government of Sweden
- For the Government of Denmark
- For the Government of Iceland
- For the Government of the Russian Federation
- For the Government of the United States of America

Appendix B
The Ilulissat declaration (2008)

Arctic Ocean conference

Ilulissat, Greenland, 27–29 May 2008

At the invitation of the Danish Minister for Foreign Affairs and the Premier of Greenland, representatives of the five coastal States bordering on the Arctic Ocean – Canada, Denmark, Norway, the Russian Federation and the United States of America – met at the political level on 28 May 2008 in Ilulissat, Greenland, to hold discussions. They adopted the following declaration:

The Arctic Ocean stands at the threshold of significant changes. Climate change and the melting of ice have a potential impact on vulnerable ecosystems, the livelihoods of local inhabitants and indigenous communities, and the potential exploitation of natural resources.

By virtue of their sovereignty, sovereign rights and jurisdiction in large areas of the Arctic Ocean the five coastal states are in a unique position to address these possibilities and challenges. In this regard, we recall that an extensive international legal framework applies to the Arctic Ocean as discussed between our representatives at the meeting in Oslo on 15 and 16 October 2007 at the level of senior officials. Notably, the law of the sea provides for important rights and obligations concerning the delineation of the outer limits of the continental shelf, the protection of the marine environment, including ice-covered areas, freedom of navigation, marine scientific research, and other uses of the sea. We remain committed to this legal framework and to the orderly settlement of any possible overlapping claims.

This framework provides a solid foundation for responsible management by the five coastal States and other users of this Ocean through national implementation and application of relevant provisions. We therefore see no need to develop a new comprehensive international legal regime to govern the Arctic Ocean. We will keep abreast of the developments in the Arctic Ocean and continue to implement appropriate measures.

The Arctic Ocean is a unique ecosystem, which the five coastal states have a stewardship role in protecting. Experience has shown how shipping disasters and subsequent pollution of the marine environment may cause irreversible disturbance of the ecological balance and major harm to the livelihoods of local inhabitants and indigenous communities. We will take steps in accordance with international law both nationally and in cooperation among the five states and other interested parties to ensure the protection and preservation of the fragile marine environment of the Arctic Ocean. In this regard we intend to work together including through the International Maritime Organization to strengthen existing measures and develop new measures to improve the safety of maritime navigation and prevent or reduce the risk of ship-based pollution in the Arctic Ocean.

The increased use of Arctic waters for tourism, shipping, research and resource development also increases the risk of accidents and therefore the need to further strengthen search and rescue capabilities and capacity around the Arctic Ocean to ensure an appropriate response from states to any accident. Cooperation, including on the sharing of information, is a prerequisite for addressing these challenges. We will work to promote safety of life at sea in the Arctic Ocean, including through bilateral and multilateral arrangements between or among relevant states.

The five coastal states currently cooperate closely in the Arctic Ocean with each other and with other interested parties. This cooperation includes the collection of scientific data concerning the continental shelf, the protection of the marine environment and other scientific research. We will work to strengthen this cooperation, which is based on mutual trust and transparency, inter alia, through timely exchange of data and analyses.

The Arctic Council and other international fora, including the Barents Euro-Arctic Council, have already taken important steps on specific issues, for example with regard to safety of navigation, search and rescue, environmental monitoring and disaster response and scientific cooperation, which are relevant also to the Arctic Ocean. The five coastal states of the Arctic Ocean will continue to contribute actively to the work of the Arctic Council and other relevant international fora.

Ilulissat, 28 May 2008

Appendix C

Observer status at the Arctic Council

Observer status in the Arctic Council is open to:

- non-arctic states
- inter-governmental and inter-parliamentary organizations, global and regional
- non-governmental organizations.

Criteria for admitting observers:

As set out in the Declaration on the Establishment of the Arctic Council and governed by the Arctic Council Rules of Procedure, observer status in the Arctic Council is open to non-Arctic States; inter-governmental and inter-parliamentary organizations, global and regional; and non-governmental organizations that the Council determines can contribute to its work.

In the determination by the Council of the general suitability of an applicant for observer status the Council will, inter alia, take into account the extent to which observers:

- Accept and support the objectives of the Arctic Council defined in the Ottawa declaration.
- Recognize Arctic States' sovereignty, sovereign rights and jurisdiction in the Arctic.
- Recognize that an extensive legal framework applies to the Arctic Ocean including, notably, the Law of the Sea, and that this framework provides a solid foundation for responsible management of this ocean.
- Respect the values, interests, culture and traditions of Arctic indigenous peoples and other Arctic inhabitants.
- Have demonstrated a political willingness as well as financial ability to contribute to the work of the Permanent Participants and other Arctic indigenous peoples.
- Have demonstrated their Arctic interests and expertise relevant to the work of the Arctic Council.
- Have demonstrated a concrete interest and ability to support the work of the Arctic Council, including through partnerships with member states and Permanent Participants bringing Arctic concerns to global decision making bodies.

Role of observers:

Decisions at all levels in the Arctic Council are the exclusive right and responsibility of the eight Arctic States with the involvement of the Permanent Participants.

- Observers shall be invited to the meetings of the Arctic Council once observer status has been granted.
- While the primary role of observers is to observe the work of the Arctic Council, observers should continue to make relevant contributions through their engagement in the Arctic Council primarily at the level of Working Groups.
- Observers may propose projects through an Arctic State or a Permanent Participant but financial contributions from observers to any given project may not exceed the financing from Arctic States, unless otherwise decided by the SAOs.
- In meetings of the Council's subsidiary bodies to which observers have been invited to participate, observers may, at the discretion of the Chair, make statements after Arctic states and Permanent Participants, present written statements, submit relevant documents and provide views on the issues under discussion. Observers may also submit written statements at Ministerial meetings.

Arctic Council Observer Manual for Subsidiary Bodies
As adopted by the Arctic Council at the
EIGHTH ARCTIC COUNCIL MINISTERIAL MEETING
Kiruna, Sweden May 15, 2013
ADDENDUM Approved by the Senior Arctic Officials at the MEETING OF THE SENIOR ARCTIC OFFICIALS
Anchorage, United States of America October 20–22, 2015
and
Portland, Maine, United States October 4, 2016

1. Background

At its Seventh Ministerial meeting in Nuuk, Greenland, the Arctic Council decided that it would publish an Observer manual to guide the Council's subsidiary bodies in relation to meeting logistics and the role played by Observers.[3]

2. Introduction

The work of the Arctic Council is to promote cooperation, coordination and interaction among the Arctic States, with the involvement of the Arctic indigenous peoples and communities of the Arctic region, on common Arctic issues, in particular issues of sustainable development and environmental protection. This work is carried out by the Council's subsidiary bodies under the guidance and direction of the Senior Arctic Officials.

Since the establishment of the Arctic Council, participation by Observers has been a valuable feature through their provision of scientific and other expertise, information and financial resources. The involvement of Observers should enhance and complement the work conducted in the Arctic Council, including the unique and critical role of Permanent Participants. In addition, Observers are encouraged to support the work of the Permanent Participants in the Arctic Council.

3. Purpose and scope

This manual reflects the provisions on Observers in the Arctic Council's founding documents, including notably the Ottawa Declaration. It complements the Rules of Procedure and guides the participation of Observers at meetings of subsidiary bodies of the Arctic Council.

The purpose of the manual is to strengthen the work carried out by the subsidiary bodies. In particular, this manual will help to guide chairs of subsidiary bodies to run meetings effectively and efficiently to ensure the exchange of information on items under discussion.

The manual is applicable to all meetings of subsidiary bodies where Observers are invited. Observers are encouraged to participate with their expertise, competence and resources primarily in Working Group meetings and projects.

In addition, this manual provides guidance for Observers' working relations with the Arctic Council Secretariat. For information purposes a copy of the Rules of Procedure is appended as Appendix 1 to this manual.[4]

4. Definitions

4.1 Membership

Membership in the Arctic Council is limited exclusively to the Arctic States: Canada, Kingdom of Denmark, Finland, Iceland, Norway, the Russian Federation, Sweden and the United States of America (the Arctic States).

4.2 Permanent participants

The Aleut International Association, the Arctic Athabaskan Council, the Gwich'in Council International, the Inuit Circumpolar Council, the Russian Association of Indigenous Peoples of the North, and the Saami Council are Permanent Participants in the Arctic Council.

The category of Permanent Participants has been created to provide for active participation and full consultation with the Arctic indigenous representatives within the Arctic Council. This principle applies to all meetings and activities of the Arctic Council.

4.3 Observers

Observer status in the Arctic Council is open to:

- **a** non-Arctic States;
- **b** inter-governmental and inter-parliamentary organizations, global and regional;
- **c** non-governmental organizations

that the Council determines can contribute to its work.

The Observers are accredited to the Arctic Council proper. Subsidiary bodies are not authorized to accredit Observers.

Observer status continues for such time as consensus exists among Ministers. Any Observer that engages in activities which are at odds with the Ottawa Declaration or with the Rules of Procedure will have its status as an Observer suspended.

An updated list of Observers is available on the Arctic Council website.

Flags of the eight Arctic States, the six Permanent Participants, and the Arctic Council.

Where the Arctic States participating in a subsidiary body agree, the chair of the body may invite any person or organization that can contribute expertise and is able to contribute to the work of that body to participate in specific meetings. These persons or organizations do not have observer status, and subsidiary bodies should thus decide on the necessity of a given expert on a meeting-by-meeting basis.

5. Decision-making in the Arctic Council

Decisions at all levels in the Arctic Council are the exclusive right and responsibility of the eight Arctic States with the involvement of the Permanent Participants. All decisions are taken by consensus of the Arctic States.

6. The role of observers

The primary role of Observers is to observe the work of the Arctic Council. Furthermore, Observers are encouraged to continue to make relevant contributions through their engagement primarily at the level of Working Groups.

7. Guidelines for subsidiary body meetings

The different roles of Arctic States, Permanent Participants and Observers in the Arctic Council subsidiary bodies should be reflected in the practical arrangements at all meetings.

The Chair is responsible for explaining the rules of order at meetings of a subsidiary body. A copy of this manual should be made available to all delegations.

Subsidiary bodies may establish operating guidelines consistent with the Arctic Council Rules of Procedure, this manual, and other relevant Council procedures.

7.1 Invitation and registration

Observers may attend meetings and other activities of the Arctic Council, unless Senior Arctic Officials have decided otherwise. The Heads of Delegation of the Arctic States may also at any time meet privately at their discretion.

The Chair of the subsidiary body in question should invite Observers to the meeting no later than 30 days in advance. The Chair may determine the maximum number of Observers in respective delegations.

Registration to attend meetings should be done in accordance with the procedure outlined by the subsidiary body as appropriate. The Chair will keep records of participation and convey these to the Arctic Council Secretariat.

Participant delegations, with the exception of documents designated as "restricted to Arctic States and Permanent Participants".

The official report provided by the Chair or the relevant Secretariat should be made available to Observers after the meeting.[5]

7.3 Seating arrangements

At meetings of subsidiary bodies Arctic States and Permanent Participant delegations, as a rule, are seated at the main table. The Chair should ensure that there is adequate seating for Observers adjacent to the main table.

Participants from sub-units of the same Observer delegation should sit together as one delegation and will be recognised as such.

Observers may not assign or designate another entity or organization to represent them at a meeting.

7.4 Discussion, statements and speaking order

The Chair should make every effort to ensure that discussions on agenda items proceed in a methodical and regular manner that encourages the exchange of information.

Observers may, at the discretion of the Chair, make statements, present written statements, submit relevant documents and provide views on the issues under discussion.

On any agenda item under discussion, the Chair should ensure that speakers from Arctic State and Permanent Participant delegations have first had an opportunity to intervene and discuss the agenda item, before considering opening the discussion to further interventions from all delegations to the meeting, including speakers from Observer delegations.

During the discussion of any matter, a representative of an Arctic State or Permanent Participant may rise to a point of order and the point of order shall be decided immediately by the Chair.[6]

7.5 Projects

Observers may propose projects through an Arctic State or a Permanent Participant but the total financial contributions from all Observers to any given project may not exceed the financing from Arctic States, unless otherwise decided by the Senior Arctic Officials.[7]

7.6 Costs

Observers are responsible for all costs associated with their attendance at a meeting of a subsidiary body.

7.7 Communications

While Arctic Council bodies are in session participants may not comment publicly on agenda items under discussion.

7.8 Other

If an Observer delegation does not respect the guidelines outlined in this manual, the Chair, after consulting with the Heads of Delegations for the Arctic States and Permanent Participants, may ask the delegation to leave the meeting. The Chair will inform the Chair of the Senior Arctic Officials accordingly.

8. Arctic Council Secretariat

The priority of the Arctic Council Secretariat is to provide services to Arctic States and Permanent Participants. It may also assist Observers in their participation by communicating information about meetings and other activities, distributing documents, as appropriate, and other assistance the Director decides to provide, in accordance with its Terms of Reference.

9. Amendment of this manual

This manual may be amended by a decision of the Senior Arctic Officials.

Addendum to the observer manual for subsidiary bodies

The following guidelines direct the engagement of Arctic Council Observers in Arctic Council subsidiary bodies[8] in accordance with the Arctic Council Rules of Procedure[9] and the Observer Manual for Subsidiary Bodies. These guidelines provide further specification for the participation of Observers, facilitating relevant and meaningful contributions to the work of the Arctic Council, where applicable.

1) Intersessional communication

As provided for in Rule 37 of the Arctic Council Rules of Procedure and Section 7.1 of the Observer Manual for Subsidiary Bodies, ahead of a meeting of an Arctic Council subsidiary body, all Observers should receive the meeting invitation no later than 30 days in advance.

As provided for in Section 7.2 of the Observer Manual for Subsidiary Bodies, ahead of a meeting of an Arctic Council subsidiary body, Observers admitted to a meeting should receive meeting documents, including a final agenda, no later than 30 days before the specific meeting, other than those documents designated as "restricted to Arctic States and Permanent Participants."

Where applicable, and to the extent possible, Observers should also receive an updated list of ongoing projects, including projects in development and points-of-contact.

2) Meeting participation

Further to Rule 38 of the Arctic Council Rules of Procedure and as provided for in Section 7.4 in the Observer Manual for Subsidiary Bodies, during a meeting of an Arctic Council subsidiary body, the Chair should make every effort to provide interested Observers with an opportunity to:

- Make statements after Arctic States and Permanent Participants, and take part in discussion concerning projects to which the Observer intends to contribute or has contributed;
- Present written statements;
- Submit relevant documents.

3) Project contributions

Observers are invited to contribute to the body of work of Arctic Council subsidiary bodies through:

- Project proposals, through an Arctic State or a Permanent Participant in a collaborative manner;
- Views expressed on projects under development, including Working Group work plans;
- In-kind contributions to existing and developing projects, such as expert involvement and support;
- Financial contributions to existing and developing projects, such as direct project funding (not to exceed the financing from Arctic States, unless otherwise decided by the Senior Arctic Officials)[10] in a transparent manner;
- Hosting of project-specific, expert-level workshops or gatherings, as approved by the SAOs on a case-by-case basis.[11]

Appendix D

Agreement to prevent unregulated high seas fisheries in the central Arctic Ocean (2018)

The Parties to this Agreement,

RECOGNIZING that until recently ice has generally covered the high seas portion of the central Arctic Ocean on a year-round basis, which has made fishing in those waters impossible, but that ice coverage in that area has diminished in recent years;

ACKNOWLEDGING that, while the central Arctic Ocean ecosystems have been relatively unexposed to human activities, those ecosystems are changing due to climate change and other phenomena, and that the effects of these changes are not well understood;

RECOGNIZING the crucial role of healthy and sustainable marine ecosystems and fisheries for food and nutrition;

RECOGNIZING the special responsibilities and special interests of the central Arctic Ocean coastal States in relation to the conservation and sustainable management of fish stocks in the central Arctic Ocean;

NOTING IN THIS REGARD the initiative of the central Arctic Ocean coastal States as reflected in the *Declaration Concerning the Prevention of Unregulated High Seas Fishing in the Central Arctic Ocean* signed on 16 July 2015;

RECALLING the principles and provisions of treaties and other international instruments relating to marine fisheries that already apply to the high seas portion of the central Arctic Ocean, including those contained in:

- the *United Nations Convention on the Law of the Sea* of 10 December 1982 ("the Convention");
- the *Agreement for the Implementation of the Provisions of the United Nations Convention on the Law of the Sea of 10 December 1982 relating to the Conservation and Management of Straddling Fish Stocks and Highly Migratory Fish Stocks* of 4 August 1995 ("the 1995 Agreement"); and
- the 1995 Code of Conduct for Responsible Fisheries and other relevant instruments adopted by the Food and Agriculture Organization of the United Nations;

UNDERLINING the importance of ensuring cooperation and coordination between the Parties and the North-East Atlantic Fisheries Commission, which has competence to adopt conservation and management measures in part of the high seas portion of the central Arctic Ocean, and other relevant mechanisms for fisheries management that are established and operated in accordance with international law, as well as with relevant international bodies and programs;

BELIEVING that commercial fishing is unlikely to become viable in the high seas portion of the central Arctic Ocean in the near future and that it is therefore premature under current circumstances to establish any additional regional or subregional fisheries management organizations or arrangements for the high seas portion of the central Arctic Ocean;

DESIRING, consistent with the precautionary approach, to prevent the start of unregulated fishing in the high seas portion of the central Arctic Ocean while keeping under regular review the need for additional conservation and management measures;

RECALLING the 2007 United Nations Declaration on the Rights of Indigenous Peoples;

RECOGNIZING the interests of Arctic residents, including Arctic indigenous peoples, in the long-term conservation and sustainable use of living marine resources and in healthy marine ecosystems in the Arctic Ocean and underlining the importance of involving them and their communities; and

DESIRING to promote the use of both scientific knowledge and indigenous and local knowledge of the living marine resources of the Arctic Ocean and the ecosystems in which they occur as a basis for fisheries conservation and management in the high seas portion of the central Arctic Ocean,

HAVE AGREED as follows:

Article 1
Use of terms

For the purposes of this Agreement:

1 "Agreement Area" means the single high seas portion of the central Arctic Ocean that is surrounded by waters within which Canada, the Kingdom of Denmark in respect of Greenland, the Kingdom of Norway, the Russian Federation and the United States of America exercise fisheries jurisdiction;

2 "fish" means species of fish, molluscs and crustaceans except those belonging to sedentary species as defined in Article 77 of the Convention;

3 "fishing" means searching for, attracting, locating, catching, taking or harvesting fish or any activity that can reasonably be expected to result in the attracting, locating, catching, taking or harvesting of fish;

4 "commercial fishing" means fishing for commercial purposes;

5 "exploratory fishing" means fishing for the purpose of assessing the sustainability and feasibility of future commercial fisheries by contributing to scientific data relating to such fisheries;
6 "vessel" means any vessel used for, equipped to be used for, or intended to be used for fishing.

Article 2
Objective of this Agreement

The objective of this Agreement is to prevent unregulated fishing in the high seas portion of the central Arctic Ocean through the application of precautionary conservation and management measures as part of a long-term strategy to safeguard healthy marine ecosystems and to ensure the conservation and sustainable use of fish stocks.

Article 3
Interim conservation and management measures concerning fishing

1 Each Party shall authorize vessels entitled to fly its flag to conduct commercial fishing in the Agreement Area only pursuant to:

   1 conservation and management measures for the sustainable management of fish stocks adopted by one or more regional or subregional fisheries management organizations or arrangements, that have been or may be established and are operated in accordance with international law to manage such fishing in accordance with recognized international standards; or
   2 interim conservation and management measures that may be established by the Parties pursuant to Article 5, paragraph l(c)(ii).

2 The Parties are encouraged to conduct scientific research under the framework of the Joint Program of Scientific Research and Monitoring established pursuant to Article 4 and under their respective national scientific programs.
3 A Party may authorize vessels entitled to fly its flag to carry out exploratory fishing in the Agreement Area only pursuant to conservation and management measures established by the Parties on the basis of Article 5, paragraph l (d).
4 The Parties shall ensure that their scientific research activities involving the catching of fish in the Agreement Area do not undermine the prevention of unregulated commercial and exploratory fishing and the protection of healthy marine ecosystems. The Parties are encouraged to inform each other about their plans for authorizing such scientific research activities.
5 The Parties shall ensure compliance with the interim measures established by this Article, and with any additional or different interim measures they may establish pursuant to Article 5, paragraph l(c).

6 Consistent with Article 7 of the 1995 Agreement, coastal States Parties and other Parties shall cooperate to ensure the compatibility of conservation and management measures for fish stocks that occur in areas both within and beyond national jurisdiction in the central Arctic Ocean in order to ensure conservation and management of those stocks in their entirety.
7 Other than as provided in paragraph 4 above, nothing in this Agreement shall be interpreted to restrict the entitlements of Parties in relation to marine scientific research as reflected in the Convention.

Article 4
Joint Program of Scientific Research and Monitoring

1 The Parties shall facilitate cooperation in scientific activities with the goal of increasing knowledge of the living marine resources of the central Arctic Ocean and the ecosystems in which they occur.
2 The Parties agree to establish, within two years of the entry into force of this Agreement, a Joint Program of Scientific Research and Monitoring with the aim of improving their understanding of the ecosystems of the Agreement Area and, in particular, of determining whether fish stocks might exist in the Agreement Area now or in the future that could be harvested on a sustainable basis and the possible impacts of such fisheries on the ecosystems of the Agreement Area.
3 The Parties shall guide the development, coordination and implementation of the Joint Program of Scientific Research and Monitoring.
4 The Parties shall ensure that the Joint Program of Scientific Research and Monitoring takes into account the work of relevant scientific and technical organizations, bodies and programs, as well as indigenous and local knowledge.
5 As part of the Joint Program of Scientific Research and Monitoring, the Parties shall adopt, within two years of the entry into force of this Agreement, a data sharing protocol and shall share relevant data, directly or through relevant scientific and technical organizations, bodies and programs, in accordance with that protocol.
6 The Parties shall hold joint scientific meetings, in person or otherwise, at least every two years and at least two months in advance of the meetings of the Parties that take place pursuant to Article 5 to present the results of their research, to review the best available scientific information, and to provide timely scientific advice to meetings of the Parties. The Parties shall adopt, within two years of the entry into force of this Agreement, terms of reference and other procedures for the functioning of the joint scientific meetings.

Article 5
Review and further implementation

1 The Parties shall meet every two years or more frequently if they so decide. During their meetings, the Parties shall, *inter alia:*

1 review implementation of this Agreement and, when appropriate, consider any issues relating to the duration of this Agreement in accordance with Article 13, paragraph 2;
2 review all available scientific information developed through the Joint Program of Scientific Research and Monitoring, from the national scientific programs, and from any other relevant sources, including indigenous and local knowledge;
3 on the basis of the scientific information derived from the Joint Program of Scientific Research and Monitoring, from the national scientific programs, and from other relevant sources, and taking into account relevant fisheries management and ecosystem considerations including the precautionary approach and potential adverse impacts of fishing on the ecosystems, consider, *inter alia*, whether the distribution, migration and abundance of fish in the Agreement Area would support a sustainable commercial fishery and, on that basis, determine:

    1 whether to commence negotiations to establish one or more additional regional or subregional fisheries management organizations or arrangements for managing fishing in the Agreement Area, and
    2 whether, once negotiations have commenced pursuant to subparagraph (i) above and once the Parties have agreed on mechanisms to ensure the sustainability of fish stocks, to establish additional or different interim conservation and management measures in respect of those stocks in the Agreement Area;

4 establish, within three years of the entry into force of this Agreement, conservation and management measures for exploratory fishing in the Agreement Area. The Parties may amend such measures from time to time. These measures shall provide, *inter alia*, that:

    1 exploratory fishing shall not undermine the objective of this Agreement,
    2 exploratory fishing shall be limited in duration, scope and scale to minimize impacts on fish stocks and ecosystems and shall be subject to standard requirements set forth in the data sharing protocol adopted in accordance with Article 4, paragraph 5,
    3 a Party may authorize exploratory fishing only on the basis of sound scientific research and when it is consistent with the Joint Program of Scientific Research and Monitoring and its own national scientific program(s),
    4 a Party may authorize exploratory fishing only after it has notified the other Parties of its plans for such fishing and it has provided other Parties an opportunity to comment on those plans, and
    5 a Party must adequately monitor any exploratory fishing that it has authorized and report the results of such fishing to the other Parties.

2 To promote implementation of this Agreement, including with respect to the Joint Program of Scientific Research and Monitoring and other activities undertaken pursuant to Article 4, the Parties may form committees or similar bodies in which representatives of Arctic communities, including Arctic indigenous peoples, may participate.

Article 6
Decision-making

1 Decisions of the Parties on questions of procedure shall be taken by a majority of the Parties casting affirmative or negative votes.
2 Decisions of the Parties on questions of substance shall be taken by consensus. For the purpose of this Agreement, "consensus" means the absence of any formal objection made at the time the decision was taken.
3 A question shall be deemed to be of substance if any Party considers it to be of substance.

Article 7
Dispute settlement

The provisions relating to the settlement of disputes set forth in Part VIII of the 1995 Agreement apply, *mutatis mutandis*, to any dispute between Parties relating to the interpretation or application of this Agreement, whether or not they are also Parties to the 1995 Agreement.

Article 8
Non-parties

1 The Parties shall encourage non-parties to this Agreement to take measures that are consistent with the provisions of this Agreement.
2 The Parties shall take measures consistent with international law to deter the activities of vessels entitled to fly the flags of non-parties that undermine the effective implementation of this Agreement.

Article 9
Signature

1 This Agreement shall be open for signature at Ilulissat on 3 October 2018 by Canada, the People's Republic of China, the Kingdom of Denmark in respect of the Faroe Islands and Greenland, Iceland, Japan, the Republic of Korea, the Kingdom of Norway, the Russian Federation, the United States of America and the European Union and shall remain open for signature for 12 months following that date.
2 For signatories to this Agreement, this Agreement shall remain open for ratification, acceptance or approval at any time.

Article 10
Accession

1 For the States listed in Article 9, paragraph 1 that have not signed this Agreement, and for the European Union if it has not signed this Agreement, this Agreement shall remain open for accession at any time.
2 After the entry into force of this Agreement, the Parties may invite other States with a real interest to accede to this Agreement.

Article 11
Entry into force

1 This Agreement shall enter into force 30 days after the date of receipt by the depositary of all instruments of ratification, acceptance, or approval of, or accession to, this Agreement by those States and the European Union listed in Article 9, paragraph 1.
2 After entry into force of this Agreement, it shall enter into force for each State invited to accede pursuant to Article 10, paragraph 2 that has deposited an instrument of accession 30 days after the date of deposit of that instrument.

Article 12
Withdrawal

A Party may withdraw from this Agreement at any time by sending written notification of its withdrawal to the depositary through diplomatic channels, specifying the effective date of its withdrawal, which shall be at least six months after the date of notification. Withdrawal from this Agreement shall not affect its application among the remaining Parties or the duty of the withdrawing Party to fulfill any obligation in this Agreement to which it otherwise would be subject under international law independently of this Agreement.

Article 13
Duration of this Agreement

1 This Agreement shall remain in force for an initial period of 16 years following its entry into force.
2 Following the expiration of the initial period specified in paragraph 1 above, this Agreement shall remain in force for successive five-year extension period(s) unless any Party:

    1 presents a formal objection to an extension of this Agreement at the last meeting of the Parties that takes place prior to expiration of the initial period or any subsequent extension period; or
    2 sends a formal objection to an extension to the depositary in writing no later than six months prior to the expiration of the respective period.

3 The Parties shall provide for an effective transition between this Agreement and any potential new agreement establishing an additional regional or subregional fisheries management organization or arrangement for managing fishing in the Agreement Area so as to safeguard healthy marine ecosystems and ensure the conservation and sustainable use of fish stocks in the Agreement Area.

Article 14
Relation to other Agreements

1 The Parties recognize that they are and will continue to be bound by their obligations under relevant provisions of international law, including those reflected in the Convention and the 1995 Agreement, and recognize the importance of continuing to cooperate in fulfilling those obligations even in the event that this Agreement expires or is terminated in the absence of any agreement establishing an additional regional or subregional fisheries management organization or arrangement for managing fishing in the Agreement Area.
2 Nothing in this Agreement shall prejudice the positions of any Party with respect to its rights and obligations under international agreements and its positions with respect to any question relating to the law of the sea, including with respect to any position relating to the exercise of rights and jurisdiction in the Arctic Ocean.
3 Nothing in this Agreement shall prejudice the rights, jurisdiction and duties of any Party under relevant provisions of international law as reflected in the Convention or the 1995 Agreement, including the right to propose the commencement of negotiations on the establishment of one or more additional regional or subregional fisheries management organizations or arrangements for the Agreement Area.
4 This Agreement shall not alter the rights and obligations of any Party that arise from other agreements compatible with this Agreement and that do not affect the enjoyment by other Parties of their rights or the performance of their obligations under this Agreement. This Agreement shall neither undermine nor conflict with the role and mandate of any existing international mechanism relating to fisheries management.

Article 15
Depositary

1 The Government of Canada shall be the depositary for this Agreement.
2 Instruments of ratification, acceptance, approval or accession shall be deposited with the depositary.
3 The depositary shall inform all signatories and all Parties of the deposit of all instruments of ratification, acceptance, approval or accession and perform such other functions as are provided for in the 1969 *Vienna Convention on the Law of Treaties*.

DONE at Ilulissat on this 3rd day of October 2018, in a single original, in the Chinese, English, French and Russian languages, each text being equally authentic.

Notes

1 The Arctic Council should not deal with matters related to military security.
2 The use of the term "peoples" in this declaration shall not be construed as having any implications as regard the rights which may attach to the term under international law.
3 Relevant background documents for this manual are the Ottawa Declaration and the Arctic Council Rules of Procedure, in particular Annex 2.
4 This applies only to the electronic version of the Observer Manual.
5 See also Addendum, p. 11 "Intersessional Communication"
6 See also Addendum, p. 12 "Meeting Participation"
7 See also Addendum, p. 13 "Project Contributions"
8 The subsidiary bodies of the Arctic Council are: 1) Working Groups; 2) Task Forces; 3) Expert Groups; 4) such other subsidiary bodies the Arctic Council may create.
9 Arctic Council Rules of Procedure, as revised by the Arctic Council at the Eighth Arctic Council Ministerial Meeting in Kiruna, Sweden, on May 15, 2013.
10 If a Working Group or other subsidiary body wishes to request an exception to Article 7.5 of the Observer Manual for Subsidiary Bodies, it should, with the support of at least one Senior Arctic Official or one Permanent Participant, submit the request in writing to the Senior Arctic Officials via a communication to the Chair of the Senior Arctic Official, along with a justification not to exceed two pages. The justification should:

   - Indicate which Senior Arctic Official(s) and/or Permanent Participant(s) support the request;
   - Briefly describe the nature of the project;
   - Estimate the overall cost of the project, the amount of anticipated financing from Arctic Council Member States and the amount being offered by Observers;
   - Identify the Observer(s) that wishes to contribute financing; and
   - Briefly explain why the Senior Arctic Officials should grant an exception to Article 7.5 with respect to the project in question.

   The Senior Arctic Officials should decide upon each such request within 30 days of its submission, either at a meeting of the Senior Arctic Officials, by teleconference, or by written communication.
11 If a Working Group or other subsidiary body wishes to accept the offer from an Observer to host a gathering, it should, with the support of at least one Senior Arctic Official or one Permanent Participant, submit the offer in writing to the Senior Arctic Officials via a communication to the Chair of the Senior Arctic Officials, along with a brief justification. The justification should:

   - Indicate which Senior Arctic Official(s) and/or Permanent Participant(s) support the offer;
   - Briefly describe the nature of the proposed gathering;
   - Indicate the Observer's interest in hosting the gathering.

   When assessing offers to host such events, subsidiary bodies and SAOs should consider the following objective criteria:

   - Observer capacity to provide relevant subject-matter expertise;
   - Observer capacity to provide a unique value-added contribution to the work of the subsidiary body by hosting the event;
   - Observer hosting as a percentage of overall project or initiative funding.

   The Senior Arctic Officials should decide upon each such request within 30 days of its submission, either at a meeting of the Senior Arctic Officials, by teleconference, or by written communication.

Index